For Bob Cooper
~ with Best Wishes,
July 2018.

Peter L.

see near foot of PM0.
For Blanchard ref, see
P.108

CLASSICAL
THEORY OF STRUCTURES

CLASSICAL
THEORY OF STRUCTURES
BASED ON THE DIFFERENTIAL
EQUATION

P. G. LOWE

Fellow of Clare College and Lecturer in
Engineering, University of Cambridge

CAMBRIDGE
AT THE UNIVERSITY PRESS
1971

Published by the Syndics of the Cambridge University Press
Bentley House, 200 Euston Road, London NW1 2DB
American Branch: 32 East 57th Street, New York, N.Y.10022

© Cambridge University Press 1971

Library of Congress Catalogue Card Number: 70–142960

ISBN 0 521 08089 4

Printed in Great Britain
at the University Printing House, Cambridge
(Brooke Crutchley, University Printer)

Ki oku matua

CONTENTS

PREFACE   *page* xi

O   PRELIMINARIES   1
    0.0   Introduction   1
    0.1   Linear ordinary differential equations   1
    0.2   Continuous particular integrals: variation of parameters   3
    0.3   Discontinuous particular integrals: the 【 ... 】 notation   4
    0.4   The governing differential equation   7
    0.5   Boundary value problems   7
    0.6   Dirac delta and Heaviside step functions   9
    0.7   Units and dimensions   9
    0.8   References   10

I   THE BEAM   11
    1.0   Physical background   11
    1.1   The fundamental equations: rigid–plastic beams   13
    1.2   The fundamental solutions: rigid–plastic beams   15
    1.3   An application: rigid–plastic beam   16
    1.4   An integral formula: rigid–plastic beams   18
    1.5   The limit theorems   19
    1.6   Elastic and elastic–plastic beams   21
    1.7   The fundamental equations: elastic and elastic–plastic
          beams   22
    1.8   The fundamental solution: elastic and elastic–plastic
          beams   23
    1.9   Discontinuous solutions: force discontinuities; elastic
          beams   25
    1.10 Discontinuous solutions: displacement discontinuities;
          elastic–plastic beams   28
    1.11 Summary of discontinuous particular integrals   32
    1.12 Further discontinuous solutions: simultaneous
          discontinuities   32
    1.13 Discontinuous solutions: two further elastic examples   34
    1.14 Continuous variations of section: member properties   36
    1.15 Discontinuous solutions: a bridge problem; elastic   37
    1.16 Symmetry and evenness   42

vii

Contents

1.17 Conclusion   *page* 42

1.18 Further exercises   43

1.19 Solutions and comments on exercises   44

1.20 References   45

2   PLANE FRAMEWORKS   47

2.0   Introduction   47

2.1   Topology of the structure: statical indeterminacy and kinematic freedom   48

PART I SIMPLE FRAMEWORKS

2.2   Simple frameworks I: rigid–plastic   51

2.3   The fundamental equations: rigid–plastic   51

2.4   The fundamental solutions: rigid–plastic   52

2.5   A rigid–plastic example: coordinate systems   52

2.6   Simple frameworks II: elastic   55

2.7   The fundamental equations: elastic   55

2.8   The fundamental solutions: discontinuous particular integrals   55

2.9   Simultaneous discontinuities: joints: elastic frames   56

2.10 Elastic frames with no joint displacements: I simple portal   57

2.11 Elastic frames with no joint displacements: II the mill frame   58

2.12 Frames with kinematic freedoms I   61

2.13 Frames with kinematic freedoms II: overall equilibrium equations   63

2.14 Simple frameworks III: elastic–plastic   68

2.15 More extensive frameworks: two storey elastic: I symmetric loading   72

2.16 More extensive frameworks: two storey elastic: II antisymmetric loading   75

2.17 Conclusion to discussion of simple frameworks   77

2.18 Further exercises   78

2.19 Solutions to exercises   81

PART II THE REGULAR MULTI-STOREY FRAME

2.20 Introduction   90

2.21 The basic problems: the equations   91

Contents

2.22	Equation solution	*page* 96
2.23	The displacements	100
2.24	Discrete and continuous formulations	103
2.25	An example: exercises	103
2.26	References	110

3 THE COLUMN — 112

3.0	Introduction	112
3.1	The fundamental equations and solutions	113
3.2	The physical phenomena: buckling and deflexion problems	115
3.3	The single column length: Euler theory: homogeneous problem	116
3.4	The single column length: non-homogeneous problems	119
3.5	The column: discontinuous solutions	121
3.6	The single column length: a discontinuous solution	123
3.7	Discontinuous particular integrals: further features	124
3.8	An example	127
3.9	Column stiffness problems: stability functions	128
3.10	Column stiffness: some applications of stability functions	132
3.11	The tie: increased stability	134
3.12	Inelastic column theory	134
3.13	Conclusions	136
3.14	Summary of discontinuous particular integrals for the column	137
3.15	Exercises	137
3.16	References	139

4 FURTHER TOPICS IN ONE DIMENSIONAL STRAIGHT MEMBERED STRUCTURES — 142

4.0	Introduction	142
4.1	Out-of-plane loading: grid frameworks	143
4.2	Example of elastic bending and twisting	145
4.3	Rigid-jointed space frameworks	148
4.4	Initial stress problems: I temperature stresses	152
4.5	Initial stress problems: II prestress	153
4.6	Braced frameworks	156
4.7	Composite beams: shear connexion and slip	160

Contents

4.8	The cable: small dip, inextensional theory	*page* 163
4.9	Conclusions and a look ahead	167
4.10	References	168

5 FOUNDATIONS OF BEAM THEORY AND
APPROXIMATIONS 169

5.0	Introduction	169
5.1	Functions and moments	169
5.2	Some operations with moments	171
5.3	The approximation procedure	171
5.4	The elasticity equations	172
5.5	The foundations of beam theory	173
5.6	A second order beam theory: shear deformation	176
5.7	Approximate solution of differential equations by use of moments	179
5.8	Some further features of approximation procedures	181
5.9	References	183

6 A MAXIMUM/MINIMUM THEOREM FOR FRAMEWORKS 184

6.0	Introduction	184
6.1	The theorem: two-dimensional version	185
6.2	Proof of the theorem	186
6.3	An example	188
6.4	An experimental result	193
6.5	Design implications	196
6.6	An alternative treatment of the theorem: matrices	200
6.7	Conclusion	202
6.8	References	202

APPENDIX OF TABLES	204
INDEX	213

GENERAL NOTE ON FIGURES

A number of structures in the figures are shown in an initial undeformed and a final deformed shape. In these cases the undeformed shape is shown by a broken line, the deformed by a full line. The deformation has also been exaggerated with use of a larger transverse than axial scale of distance. This has been done to make clearer the type of deformation experienced, but it should be remembered that all deformations are assumed small.

PREFACE

The subject of the theory of structures we shall take to mean the study of the static strength, stiffness and stability of engineering constructions.

Traditionally, courses in the theory of structures, in British and Commonwealth Universities at least, have been largely given over to the study of a limited range of structural forms – the beam, the column and the framework and with a civil engineering bias to the treatment. Many of the long established methods developed for solution of problems presuppose such an application. The major emphasis is on establishing what forces are acting and these, in civil engineering, usually govern the design. The associated *displacements* are seldom accorded an equal status with the forces, except when the computations are being arranged for a digital computer. In mechanical engineering structural problems, on the other hand, very often the reverse is true and displacements tend to govern the design.

The presentation of our subject here treats mainly of members in *bending*. The method proposed ensures that the forces and displacements in any structural member or system are treated with something approaching equality and this circumstance, it seems to me, allows a much tidier presentation of the subject. A single process is advanced for all the types of problem encountered and this unification it is believed provides an advantage in teaching the subject over a program which embraces for example Castigliano's theorems, column analogy, moment distribution, influence coefficients and slope deflexion methods, as the methods for dealing with rigid jointed *elastic* frameworks. Such a program, besides explaining the content of the method, must then go on to explain in what situations each method can best be applied. To counter these claims it could be argued that the method here proposed is not 'optimum' in any situation; for example, it is not as efficient a computational device as the column analogy for dealing with single ring elastic structures. I think a sufficient answer is merely to note that the method proposed is well able to cope with this type of problem and, in addition, many others with which the column analogy is powerless to deal. The present approach it will be argued is a viable method, *whether by hand or digital machine*, for dealing with a wide range of elastic and plastic problems met with in practice and is admittedly not specifically adapted to dealing with a particular class of structure.

But there is a further and deeper reason for the present choice of method and the present emphasis on both forces *and* displacements. This is that the real interest and goal in many situations in practice is to examine, and hopefully solve, the *two dimensional* structural problems of plates and shells – the floor slabs, water tanks, motor bodies, aircraft structures and ship hulls of practice, rather than just the *one dimensional* beam, column and framework. It is suggested that the present approach will allow more development of skill and the opportunity to retain more of pure *method* when we move from dealing with the one dimensional beam and column to the two dimensional plate and shell than is the case with traditional framework methods.

In the field of plates and shells, whether the idealization is to elastic or inelastic material behaviour, most problems cannot be solved in closed form. We must resort to approximate methods and the most promising methods available all have in common the need to discuss the *displaced form* at the outset. Only subsequently, if at all, are the internal force systems discussed. Clearly then, systematic discussion of displacements should not be thrust into the background in first-degree courses.

An important ingredient in the present treatment of one dimensional problems is the systematic exploitation of *discontinuity concepts of the under-lying differential equation*. A typical discontinuity in a beam problem is a 'point' transverse load, really acting over a short length of the beam but idealized to act at a point. The Macaulay 'bracket' is a device known to many and used to deal with such a discontinuity in the shear force. There are other frequently occurring discontinuities, for example the plastic hinge in a structure near to or at collapse. This is a discontinuity in the slope of the member, a discontinuity in a displacement variable. Looked at from this point of view, it is clear that much progress can be made not only in bringing together the elastic and the plastic streams in framework analysis but also in analysing frameworks in general, with the *expectation* that discontinuities will be the rule rather than the exception. This amounts to a near reversal of present trends in framework analysis methods.

Thus a rigid joint between a beam and a column in what follows will itself appear as a particular sort of discontinuity, a typically simultaneous discontinuity in shear and displacement. In this way, all quantities of interest, namely the basic four of transverse displacement, slope, shear force and bending moment in a plane straight member, will be seen to be *continuous* everywhere, except at points where they are shown to be *discontinuous* – for example at points of load application, direction change or at a plastic hinge. Then the transition from a straight beam to a framework or from an elastic to an elastic–plastic material can be most easily made and the differences are readily identifiable.

In the present treatment of structures through discontinuity principles all examples are discussed as applied to hand computation. However discontinuity methods as a computational procedure are well adapted to being programmed for a digital computer and it is hoped to expand upon this topic in a later volume. Again, only straight membered structures are dealt with, except for one example in the final chapter. Curved membered structures will perhaps be discussed in a later volume.

To sum up, the basic difference between the present approach to developing the structural theory relevant to frameworks in bending and most other accounts of the subject is that we choose to emphasize that the notions used to describe the structural behaviour are basically differential equations and we aim to exploit this fact to the full. I should hasten to add however that there is next to no mathematics needed for this exploitation, merely a determination to view the subject from a particular direction.

It is, I believe, a fair generalization to say that most other modern accounts of the subject emphasize and exploit notions of work and energy, and especially the virtual work equation as a description of equilibrium. While it is true that the one approach can be transformed into the other, in practice this cannot be done without some difficulty and would be wasteful of effort if the aim is, basically, to equip the reader to solve problems. This end can be achieved along either path. But because of what seems to me an imbalance in the available literature, where energy treatments clearly predominate, I have felt justified in giving what I hope is a connected account of the subject without the use of energy considerations. For some classes of problem it is certainly true that energy methods lead to a neater and quicker solution, and the converse is also true. I believe however that there are some educational and practical reasons to justify a treatment which avoids energy concepts. But the reader should not neglect the energy principles part of his education and I have taken pains to indicate where such developments can be found and broadly how they link up with the present development of the subject.

To some the word 'classical' on the title page may seem inappropriate. My thinking here is that I am endeavouring to raise with the reader, even if only during the time he is reading these words, questions of newness and heritage. In some small way I hope to set him thinking about the historical development of this subject. For at least 100 years and possibly longer the notions which we exploit of 'discontinuity' have been contained in the accessible literature of engineering calculations and for much longer all the differential equation techniques which we use here have been available. What I trust is *new* is the *extent* to which the ideas are here exploited. In fact, the period during which the relevant theory for both the differential equations used here and the virtual work concepts used in alternative treatments was developed was the century following the middle eighteenth century. This then is my justification for use of the word 'classical'.

All the central features of the point of view adopted in the following pages have been taught to classes in Cambridge in the past few years. Inevitably though competition with other subjects and other constraints on the syllabus have meant that only the bones of the approach have shown through. Perhaps this more extended account will allow the methods to be seen in better perspective. The book is not a text in the usual sense although it is elementary in character. It is my hope, however, that it might be read by the engineer early in his career.

I have had the good fortune to be influenced by a number of teachers, employers, colleagues and friends. I would wish to mention especially Cecil Segedin, Ronald Jenkins, Hugh Tottenham, Ronald Tiffen and Jacques Heyman. They may not agree with my choice of method or the emphasis I have adopted but they have contributed in various ways to my thinking about the subject.

P. G. LOWE

Friarswood,
102 Long Road,
Cambridge
July 1970

0

PRELIMINARIES

0.0 Introduction

The basic theme running through most mathematics used in engineering is that of *continuity*. For example, those parts of the calculus used most extensively emphasize differentiability and hence continuity. This circumstance has influenced engineering computation very markedly. But in what follows we intend to exploit the theme of *discontinuity* and emphasize the unity and gains which result, since discontinuities occur naturally and frequently in application, with comparable frequency to continuities. For completeness we shall review that corner of differential equation theory we require to use. But first some pieces of notation will be explained.

It is assumed that concepts of resultant force and moment across a section in a stressed member are familiar. At a number of points in the following pages forces and moments will be denoted by vector quantities. Thus a vector force will be denoted by a single headed arrow, $\rightarrow$ directed along the line of action of the force; a vector moment will be shown as a double headed arrow, $\twoheadrightarrow$ directed along the *axis* of the moment and interpreted according to a right-hand rule.

Frequently there will be a need to refer to differential equations or relations. Rather than always writing dM/dx, the notation M' will be used without further comment. Where multiple differentiation is implied, a roman numeral or a bracketed letter will be used: thus w^{iv} means d^4w/dx^4 and $w^{(n)}$ means d^nw/dx^n. The words *ordinary differential equation* will occur frequently, and will be abbreviated to O.D.E.

Occasionally there will be a need to discuss sequences in which case it is convenient to use the notation $n = 1(1)3$ to indicate that n, the index in the sequence, begins with the value 1, ends with the value 3 and steps on in increments of one.

0.1 Linear ordinary differential equations

This section and the next two are fundamental to the later discussion and the reader should take care to understand all the points made and work all the exercises provided.

Preliminaries

The static equilibrium of a beam or column element is described by an ordinary differential equation, albeit a very simple one. Thus the beam equilibrium equation is

$$M'' = \frac{d^2 M}{dx^2} = p \tag{0.1.0}$$

where M is the bending moment (B.M.) resulting from the action of a distributed loading p. The equation is *ordinary* because there is a single *independent variable*, x. An equation such as (0.1.0) will be denoted in general by

$$L_n(u) = R, \quad \text{where} \quad L_2(..) = d^2(..)/dx^2 = (\ )''$$

in the present case and is called the *operator*; R is a known function and is to do with the loading.

The study of the forces and displacements which develop in frameworks whose members are bent under load is predominantly a study of members each described by a *fourth order* O.D.E. It is usually the case that these equations have constant coefficients. This a consequence of the majority of members used in practice being of constant section properties between support points. In contrast, the dynamics of mass-spring systems or electrical L-R-C systems, is predominantly the study of *second order* O.D.E.s.

The O.D.E.s describing the behaviour of beams, and columns in which instability is not of primary importance, are so simple as to be trivial. There is no need to use anything but the simplest knowledge of the theory of differential equations to obtain a solution. But, as will be seen in the later discussion of such problems, there is a considerable benefit available if the differential *nature* of the equations is exploited to the full.

The reason is that our entire discussion of structural behaviour through the differential equation is permeated with the need to construct suitable particular integrals for a variety of situations which, along with the complementary function, form the starting point for the computations in a given case. First then let us re-examine what is meant by a particular integral and a complementary function.

The *complementary function*, hereafter denoted by C.F., is the general solution of a given equation when the right-hand side is a zero, that is, when the equation is homogeneous. This solution will involve arbitrary constants equal in number to the order, n, of the equation. The cases here of interest are static problems and the describing equations most usually are even ordered. In these cases half the arbitrary constants are found from the *boundary conditions* at one end of the domain of integration and

2

half at the other end. Any odd ordered static problems will be treated as they occur.

The *particular integral,* abbreviated to P.I. in what follows, is any solution of the nonhomogeneous equation for a given right-hand side and contains *no* arbitrary constants. But the point to be noted is that this P.I. is not unique, since clearly any proportion of any component function which appears in the C.F. could be added to this P.I. and would in turn yield a perfectly satisfactory P.I. Note too that there is no influence from the boundary conditions on the choice of the P.I.

Central to the methods about to be described for analysing frameworks is the need to construct suitable P.I.s. These are of two types, either the conventional continuous particular integrals (P.I.) or, very frequently, discontinuous particular integrals (D.P.I.).

0.2 Continuous particular integrals: variation of parameters

In this section we review the method of variation of parameters as a method for computing conventional continuous particular integrals.

Suppose we require a P.I. for the nth ordered O.D.E.

$$L_n(w) = R(x). \tag{0.2.1}$$

Let the C.F. of (0.2.1) be

$$w = c_1 w_1 + c_2 w_2 + \ldots + c_n w_n. \tag{0.2.2}$$

Then the starting point for the method of variation of parameters is to suppose the complete solution for w to be given by

$$w = V_1 w_1 + V_2 w_2 + \ldots + V_n w_n, \tag{0.2.3}$$

where the unknown *constants* c_i in (0.2.2) have been replaced by unknown *functions* $V_i(x)$.

Assume the coefficient of $w^{(n)}$ in (0.2.1) to be unity, which is always simple to arrange, then the $V_i(x)$ can be found from the n relations:

$$\left.\begin{aligned}
V_1' w_1 + V_2' w_2 + \ldots + V_n' w_n &= 0, \\
V_1' w_1' + V_2' w_2' + \ldots + V_n' w_n' &= 0, \\
&\;\;\vdots \\
V_1' w_1^{(n-2)} + V_2' w_2^{(n-2)} + \ldots + V_n' w_n^{(n-2)} &= 0, \\
V_1' w_1^{(n-1)} + V_2' w_2^{(n-1)} + \ldots + V_n' w_n^{(n-1)} &= R(x),
\end{aligned}\right\} \tag{0.2.4}$$

where the first $(n-1)$ relations ensure that

$$w^{(r)} = V_1 w_1^{(r)} + V_2 w_2^{(r)} + \ldots + V_n w_n^{(r)} \tag{0.2.5}$$

and the *n*th relation (0.2.4) follows from substituting (0.2.5) into the equation (0.2.1).

The *n* relations (0.2.4) can be solved for the V_r' and the V_r then found by quadratures (See Note 0.3). In applications, equations of the *second* order are frequently encountered in which case

$$V_1 = -\int \frac{w_2(x)}{\Delta(w_1, w_2)} R(x)\, dx, \quad V_2 = \int \frac{w_1(x)}{\Delta(w_1, w_2)} R(x)\, dx \qquad (0.2.6)$$

where
$$\Delta(w_1, w_2) = \text{Wronskian of } w_1,\, w_2$$

$$\equiv \begin{vmatrix} w_1 & w_2 \\ w_1' & w_2' \end{vmatrix}.$$

More specifically, if $L_2(u) = u'' + k^2 u$, the 'simple harmonic' equation, then $w_1 = \sin kx$, $w_2 = \cos kx$ and $\Delta(w_1, w_2) = -1$.

Any good text on O.D.I.s can be consulted for further details but Ince (0.8) is especially recommended. The method is due to Lagrange.

0.3   Discontinuous particular integrals: the 〖...〗 notation

The task before us is to construct P.I.s which are everywhere continuous except for isolated points where they are to possess specified discontinuities. The search proceeds along the lines of first computing any continuous P.I. implied by any R.H.S. present and then exploiting the inherent arbitrariness of this choice to satisfy the *discontinuity conditions*. The process can best be understood with a specific example.

Suppose the governing differential equation (see 0.4) concerned is $ß . w^{iv} + P . w'' = p$. This is an equation encountered later, but without inquiring as to the physical meanings for the symbols appearing in it, suppose it is required to construct a D.P.I. for the situation that $p = 0$ for $x < a$ and $p = \text{const} \neq 0$ for $x > a$. This *step change* in p at $x = a$ *will be denoted by* 〖p〗$_a$ at $x = a$. Generally 〖...〗 will be used exclusively to denote 'discontinuity' or 'step change in' at some value of x which will be clear from the context. Looking back at the equation under consideration, the requirements will be

$$〖w^{(n)}(a)〗 = 0, \quad n = 0(1)3. \qquad (0.3.1)$$

With the notation $\alpha^2 ß = P$, the independent functions constituting the C.F. are seen to be $\sin \alpha x$, $\cos \alpha x$, x and a constant and, since the C.F. is continuous everywhere, the *most general* P.I. will be

$$b \sin \alpha x + c \cos \alpha x + dx + e + (px^2 / 2P). \qquad (0.3.2)$$

0.3 *Discontinuous particular integrals*

Here $px^2/2P$ is the regular, continuous P.I. and the additional terms with unknown coefficients b–e remove any arbitrariness and give just the required freedoms to satisfy the discontinuity requirements.

Now

$$w(x) = \text{C.F.} + \text{P.I.},\qquad(0.3.3)$$

and

$$[\![w(a)]\!] = [\![\text{P.I.}(a)]\!]\qquad(0.3.4)$$

since

$$[\![\text{C.F.}(a)]\!] = 0.$$

Hence the discontinuity relations (0.3.1) imply the following relations at $x = a$

$$
\begin{aligned}
[\![w]\!] &= b\sin\alpha a + c\cos\alpha a + da + e + (pa^2/2P) = 0,\\
[\![w']\!] &= \alpha b\cos\alpha a - \alpha c\sin\alpha a + d \quad + (pa/P) = 0,\\
[\![w'']\!] &= -\alpha^2(b\sin\alpha a + c\cos\alpha a) \quad\quad\; + (p/P) = 0,\\
[\![w''']\!] &= \alpha^3(-b\cos\alpha a + c\sin\alpha a) \quad\quad\quad\;\; = 0,
\end{aligned}
\qquad(0.3.5)
$$

or, in matrix form,

$$
\begin{bmatrix}
\mathscr{S} & \mathscr{C} & a & 1\\
\alpha\mathscr{C} & -\alpha\mathscr{S} & 1 & 0\\
-\alpha^2\mathscr{S} & -\alpha^2\mathscr{C} & 0 & 0\\
-\alpha^3\mathscr{C} & \alpha^3\mathscr{S} & 0 & 0
\end{bmatrix}
\begin{bmatrix} b\\ c\\ d\\ e \end{bmatrix}
= -\frac{p}{2P}
\begin{bmatrix} a^2\\ 2a\\ 2\\ 0 \end{bmatrix},
\qquad(0.3.6)
$$

where $\mathscr{S} \equiv \sin\alpha a$, $\mathscr{C} \equiv \cos\alpha a$.

Now the determinant of the coefficient matrix has the value $-\alpha^5$ and hence

$$
\begin{bmatrix} b\\ c\\ d\\ e \end{bmatrix}
= \frac{1}{\alpha^5}
\begin{bmatrix}
0 & 0 & \alpha^3\mathscr{S} & \alpha^2\mathscr{C}\\
0 & 0 & \alpha^3\mathscr{C} & -\alpha^2\mathscr{S}\\
0 & -\alpha^5 & 0 & -\alpha^3\\
-\alpha^5 & \alpha^5 a & -\alpha^3 & \alpha^3 a
\end{bmatrix}
\begin{bmatrix} a^2\\ 2a\\ 2\\ 0 \end{bmatrix}
\frac{p}{2P}.
\qquad(0.3.7)
$$

If b, c, d, e are evaluated from (0.3.7) and substituted into (0.3.2) it will be found that the required particular integral is

$$w = \frac{p}{2P\alpha^2}(\alpha^2 z^2 + 2\cos\alpha z - 2).\qquad(0.3.8)$$

Here the *notation* is $z \equiv x - a$ and the *convention to be followed is that expressions which are functions of z are identically zero for $z < 0$.*

In the following pages there will be frequent occasion to construct D.P.I.s which, combined with the ever present (continuous) C.F., will be capable of describing prescribed discontinuity conditions at stated points in the structure. In every case the derivation *can* start with the general

5

solution for the given operator and will proceed exactly as in the present case although the situation in applications will be that D.P.I.s associated with a given O.D.E. are computed once only and thereafter merely used. The D.P.I.s computed in this manner will be continuous in all derivatives except the stated one (or ones) and, since ⟦ C.F. ⟧ = 0, namely the C.F. is always continuous in the range of integration, then w, in the case considered above, will itself have the same isolated discontinuities as the D.P.I.

Notes

0.1   The full mathematical implications of allowing discontinuous behaviour in the dependent quantities are not brought out in the discussion just concluded. In particular, if a more exact interpretation of a discontinuity is sought it will be apparent that at the actual *point* of discontinuity in w', say, there will be delta function behaviour (see 0.6) in the w''. This information is exploited in any integral formulae developed from the given O.D.E. and an example of this type is included in chapter 1. But in all the main applications studied in what follows, integral formulations will be avoided and the simple interpretation of a discontinuity in a given derivative of a variable as leaving all other derivatives continuous will suffice.

0.2   Some of the aspects of discontinuity used in the present treatment are used extensively in the literature dealing with Laplace transforms (L.T.) and operational methods (O.M.). The main areas of engineering applications for these techniques are those dealing with the dynamics and especially the transient response of discrete linear systems – in fact those areas where the fundamental describing equations are linear O.D.E.'s but with *time* as the *independent variable*. These problems are all of *initial value type* (I.V.T.) whereas the present structural problems are of *boundary value type* (B.V.T.). The distinction is important because although the L.T. is the appropriate tool for dealing with O.D.E.s of I.V.T., where the independent variable is of infinite extent, such as time, it is much less useful for dealing with O.D.E.s of B.V.T. where the independent variable is of finite extent.
     However, to supplement and consolidate the present discussion of discontinuous particular integrals, reference could usefully be made to any good text dealing with the Laplace transform, or, more broadly, operational methods in general.

0.3   At some points in the further discussion the term *integration in quadratures* will be encountered. This is simply taken to mean that the integrations can be effected merely by writing integral signs on both sides of the equations and proceeding with the integration. Thus the beam equation $ßw^{iv} = p$ is completely integrable in quadratures, the column equation $ßw^{iv} + Pw'' = p$ is partially integrable in quadratures and the beam-on-elastic-foundation equation $ßw^{iv} + kw = p$ is not integrable in quadratures at all.

6

0.4   When discussing variable coefficient equations the resulting D.P.I.s are generally not expressible as functions of z, in which case the z *notation* is no longer useful although the *convention* of D.P.I. $\equiv$ o for $x < a$ clearly is.

Exercises

(1) For the equation $w^{iv} = p$, derive suitable discontinuous D.P.I.s for step changes in $w^{(n)}$, $n = $ o(1)4 in turn.

    Ans. For $[\![w^i(a)]\!] = \Delta_i$,   D.P.I. $\equiv \Delta_i z^i / i!$.

(2) For the equation $w^{iv} + \alpha^2 w'' = p$, namely the equation studied in o.2, derive suitable D.P.I.s for step changes in $w^{(n)}$, $n = $ o(1)3 in turn.

    Ans. $[\![w]\!] = \Delta$,   D.P.I. $= \Delta^\circ$; $[\![w']\!] = \theta$,   D.P.I. $= \theta z$;
      $[\![w'']\!] = M$,   D.P.I. $= M/\alpha^2 \cdot (1 - \cos \alpha z)$;
      $[\![w''']\!] = F$,   D.P.I. $= F/\alpha^3 \cdot (\alpha z - \sin \alpha z)$.

0.4   The governing differential equation

In discussing any physical problem, at the outset all the dependent physical variables of interest must be listed. In the discussion of applied statics soon to be commenced, the quantities of interest are displacements and slope changes on the one hand and forces and moments on the other.

The problem will then be formulated in terms of a set of O.D.E.s; often these equations are each of first order. It will generally be the practice to eliminate all but one of the dependent variables, which one is usually dictated by the equations themselves, to produce a single equation. If the single dependent variable has been chosen correctly, once this single equation and associated boundary conditions has been solved, all other dependent quantities can be found by *differentiation.*

If this is the case the single equation will be called the *governing differential equation* (G.D.E.). For the entire subject of stable elastic and elastic–plastic structures composed of straight members in bending, the G.D.E. is the very simple equation

$$ß w^{iv} = p,$$

where w is the transverse displacement, $ß$ the section stiffness (EI) and p the distributed transverse load/unit length.

0.5   Boundary value problems

It is very frequently the case that the G.D.E. of boundary value problems, typically problems of statics, are even ordered in the independent (space) variable. By this is meant that the only derivatives which occur in the

G.D.E. are of the form $w^{(2n)}$. It might possibly be inferred from intuitive arguments that this is to be expected.

The majority of physical situations giving rise to boundary value problems do in fact lead to even ordered G.D.E.s but there are some physical situations in which odd ordered G.D.E.s are encountered. An example is the equilibrium equation of a plane curved member in a determinate or hinge collapse condition. Clearly, since the equation is odd ordered there will be an *odd* number of arbitrary constants in the C.F. and hence the two end points of the member or range of integration cannot be treated equally. The case cited is third order and the end with two conditions can usually be easily distinguished from the end with one only.

This same arch member discussed above, but in an indeterminate condition (see chapter 1), is governed by a sixth order G.D.E. and here now there are three conditions at each end. The situation never arises that the discrepancy between numbers of boundary conditions at each end of a domain of integration is greater than one. Hence for even ordered G.D.E.s the numbers of conditions at each end will be identical.

To some extent it could be argued that the concept of a 'governing differential equation' as we have defined it is imprecise. Without discussing technicalities, we shall adopt the view point that there are evident basic physical variables of force and moment on the one hand and displacements and their derivatives, slope and curvature, on the other. We do not, for example, include the integral of the displacement over some prescribed interval as a physical variable. Provided we can agree as to what the physical dependent variables of interest are, then the G.D.E. concept is useful. Further, B.V. problems, namely problems formulated in terms of a *spatial* independent variable, are generally even ordered.

Note

0.5 In the theory of axisymmetric small displacements of elastic plates what appears to be a third order G.D.E. for the transverse displacement (w) is sometimes presented, i.e. $(1/r(rw')')' = F/D$, $(\ )' = d(\ )/dr$, and F being the shear force. This is not a G.D.E. however since F is itself an unknown. If F is eliminated using the relevant force equilibrium equation then the usual fourth order G.D.E. for w is recovered.

0.6   Dirac delta and Heaviside step functions

These two functions are used extensively in the literature of Laplace transforms and elsewhere and, although we do not use the ideas explicitly, there is a need to be aware at least of the nature of these functions.

The Dirac delta function, $\delta(a)$, is a function (fig. 0.6.0 (*a*)) which is zero everywhere except that it takes on an undefined large value over an infinitesimal range in the neighbourhood of $x = a$ and such that the integrated value through this non-zero zone is unity.

The Heaviside step function, $H(a)$ (fig. 0.6.0 (*b*)), is a function which is zero for $x < a$ and equal to unity for $x > a$ with a step change connecting the regimes. Clearly $\delta(a)$ and $H(a)$ are closely related, the former being the 'derivative' of the latter. Nevertheless, operations such as differentiation, for example, need careful investigation and in our further discussion no results of this investigation are assumed.

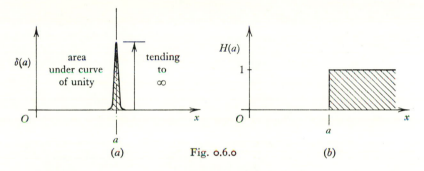

Fig. 0.6.0

An accessible account of these functions is contained in Lighthill (0.8) but any good text dealing with Laplace transforms would make satisfactory reading.

0.7   Units and dimensions

It will be assumed that the reader is familiar with the dimensions of a physical quantity in terms of mass, length and time, and will be expected to use this ability and analyse the dimensions of any symbol defined in the further development. To this end no fully numerical examples are included in the book. Instead quantities will be defined as L length units or W force units. In specific practical cases particular units can be chosen and all the results translated. It is anticipated however that either S.I. or F.P.S. units will be used in practice.

0.8 References

Ince, E. L. *Ordinary Differential Equations*, Longmans Green, London (1927). Our requirements for theory on this topic are very modest and Ince's book is very detailed. Its usefulness is in supplying background and the historical information contained is useful for forming a perspective into the subject. Modern topics like Dirac and Heaviside functions are not discussed however.

Lighthill, M. J. *Introduction to Fourier Analysis and Generalised Functions*, Cambridge University Press, London (1958). Any additional background to Dirac and Heaviside functions can be found here, if the need arises.

1

THE BEAM

1.0 Physical background

The beam is the familiar load bearing member, generally encountered with its length horizontal, under gravity loads, and spanning between support points. The primary action is one of *bending* and in this first order technical theory all other effects, such as deformation due to shear forces or instabilities, will be assumed unimportant in assessing the strength and stiffness of the beam.

The requirement in practical problems is first to obtain a value for the maximum load the beam is likely to support with safety, the *strength information*, and secondly to obtain information about the displacements produced in the member under load, the *stiffness information*. The strength information is generally the more important and for ductile materials this can be obtained from a very simple theory which assumes the beam material to be *rigid–plastic*. This model for the real material assumes the material to remain undistorted until a critical maximum moment, the fully plastic moment M_p, is reached at sufficient sections of the beam for a *mechanism* to be formed. The structure then deforms, collapses, at this maximum load.

The moment/curvature relation for a beam of such a material is as shown in fig. 1.0.0(*a*). The loading arrangement from which an experimental plot can be obtained is shown in fig. 1.0.0(*d*). However, the rigid–plastic theory is unable to provide any stiffness information. A more realistic description is the *elastic–plastic* material with a beam moment/curvature relation shown in fig. 1.0.0(*b*). With this type of material description there is a *proportional* response between moment and curvature until the maximum moment attains the full plastic value, after which this same full plastic moment is a constant moment resisting further rotation associated with a given hinge at all stages of the loading until the final hinge to produce a mechanism is formed. The load/displacement plot for a typical point on a frame consists of a series of straight line segments and, subsequent to the last hinge forming, in theory at least, large displacements and rotations may develop. In practice, the existence of a rising portion well to the right on the M/κ plot, the strain hardening

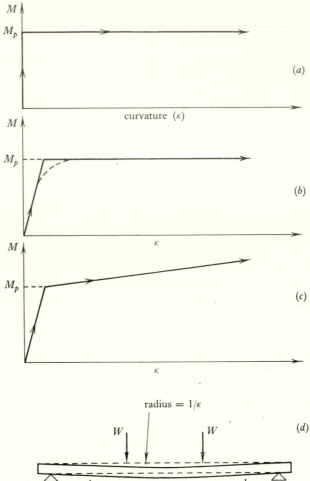

$$M = W.d, M_p = W_c.d$$

Fig. 1.0.0

zone, and geometry change effects, mean that the frame displacements are often arrested before very large displacements develop.

A further yet more sophisticated M/κ relation is shown in fig. 1.0.0(*c*): the *elastic-strain hardening* response. The analysis for this response can be developed from the methods soon to be discussed but we shall not discuss the development further here.

Experimental observations of steel or other ductile metal frames, and concrete frames suitably reinforced, give results which agree well with predictions based on the idealized moment/curvature responses shown in fig. 1.0.0 (*a*, *b*) and this is the justification for the methods of analysis about to be discussed. We should emphasize that *ductility*, that is the existence of a considerable range of curvatures for the same full plastic moment, typical especially of steels, will be assumed whenever plasticity is discussed.

Occurrence of any rotation at the full plastic moment indicates an irreversible happening. The initially straight beam will acquire some permanent bend. Strictly, irreversibility begins before M_p is reached at a section; it begins when $M > M_y$. M_y is called the *yield moment*, and M_p/M_y, defined to be the *shape factor*, has the value 1·5 for the rectangular section. For the commonly occurring H or I shape of a rolled steel section, the shape factor is nearer to unity, usually about 1·15. If $M < M_y$ everywhere then the beam is fully elastic and all effects will be fully recoverable. In this case the term *elastic* alone will be used to describe the beam response. The idealized elastic–plastic moment curvature relation here assumed presumes the shape factor to be unity, but this approximation is adequate for many practical purposes. If the shape factor is not unity then the elastic–plastic junction in the M/κ relation is rounded off as indicated by the broken line in fig. 1.0.0 (*b*). To incorporate such a refinement would unnecessarily complicate the analysis without adding any significant new feature to the mechanical behaviour.

The members studied in this chapter will be assumed to have at least one axis of symmetry, a principal axis, and any applied loads will be assumed to bend the member about an axis normal to a principal axis of the section. This is the common physical situation and should be striven for as a practical requirement. This will ensure that all displacements develop in the plane of the member and the loads. Thus beam systems can be adequately treated as plane.

1.1 The fundamental equations: rigid–plastic beams

From a consideration of the equilibrium of the beam element shown in fig. 1.1.0 there follow the relations:

$$M' = F, \quad F' = p \qquad (1.1.0)$$

where $(..)' \equiv \dfrac{\mathrm{d}(..)}{\mathrm{d}x}$ and M, F, p are respectively the internal bending moment and shear force and the external applied distributed loading.

If F is eliminated then
$$M'' = p. \qquad\qquad (1.1.1)$$

This is a fundamental equation of the present study.

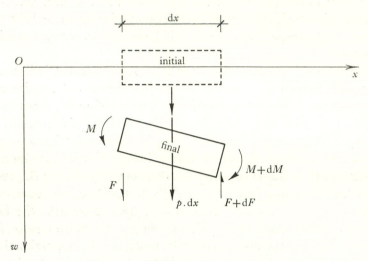

Fig. 1.1.0   Rigid–plastic beam element.

The assumed state of deformation of the structure is one of complete rigidity until a mechanism is formed, when rotation takes place at a finite number of *hinge* positions. The appearance of the structure is then one of straight segments with hinges or supports between the segments (see fig. 1.1.1(*b*)). Such a deformation is described by

$$w'' = 0, \qquad\qquad (1.1.2)$$

where $w(x)$ is the displacement measured normal to the member at x. The equation (1.1.2) is the statement of the *material properties*: the rigid–plastic requirement.

In geometrical terms a hinge is a discontinuity in the slope of a member; it is a point where the slope changes abruptly (fig. 1.1.1(*b*)). In statical terms a point transverse load is a discontinuity in the shear force F. There is therefore an evident duality between w' and F, and w and M. We shall discuss this property a little later (1.4). For simplicity it will frequently be assumed that all external loads are applied as point concentrated loads.

Then, necessarily, the only sites for hinges will be load or support points or the end points of members.

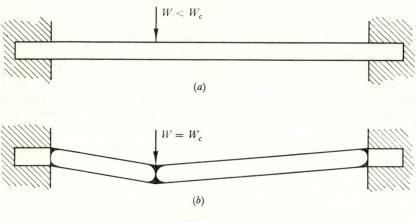

(a)

(b)

Fig. 1.1.1

1.2 The fundamental solutions: rigid–plastic beams

The equilibrium equation (1.1.1) integrates immediately to give

$$M = Ax + B + \text{P.I.} \tag{1.2.0}$$

The displacement equation (1.1.2) likewise integrates to give

$$w = ax + b + \text{P.I.} \tag{1.2.1}$$

It is important to recognize that the arbitrary constants A, B; a, b are in turn the *shear force* and *bending moment* at the chosen origin for x, and the *slope* and *displacement* at the same point.

Mention of discontinuities in slope and shear force suggests that discontinuous P.I.s may need to be added to (1.2.1) and (1.2.0), respectively. The results are trivial in the present case and are given without proof. The reader should however verify their suitability and correctness.

Clearly Wz, $z = x - a$, is a suitable D.P.I. for a downward transverse load applied at $x = a$ and θz is a suitable D.P.I. for a positive (clockwise) slope change of amount θ through a hinge at $x = a$. It should be remembered that any term with a z multiplier is omitted if z is negative! Only two other D.P.I.s are of frequent physical occurrence. They are for applied moment M at $x = a$ (positive anticlockwise) when D.P.I. $= M(z)^\circ$ and uniformly distributed load p beginning at $x = a$ when D.P.I. $= pz^2/2!$. Should the need arise other discontinuities can be quickly examined as and when they arise.

15

Note

1.1 *Sign conventions.* All the sign information needed is contained on fig. 1.1.0 and in any case of doubt this figure should be quickly sketched by the reader on the corner of his note paper. As regards the sign of a discontinuous quantity, it must be remembered that it is positive when such as to cause a positive increase in the action or displacement under consideration. Thus a concentrated applied moment will be an anticlockwise moment in order that the *internal* bending moment in the member be increased positively. A hinge rotation θ will be positive when the portion of the beam to the right is rotated clockwise with respect to that to the left, since this produces a positive increase in w'. This discontinuous quantity sign information is summarized on fig. 1.1.2.

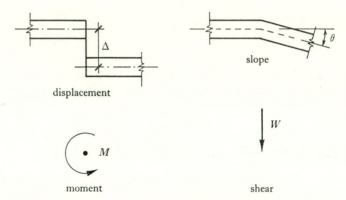

displacement   slope

moment   shear

Fig. 1.1.2 Positive discontinuities.

1.3 An application: rigid–plastic beam

Let us now compute the collapse load for the beam shown in fig. 1.3.0.

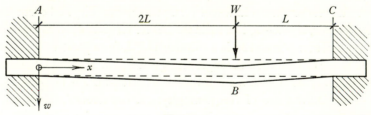

Fig. 1.3.0

By inspection it is seen that three hinges are needed to reduce the structure to being a mechanism. In this particular case there are only three possible hinge sites, A, B, and C and hence the problem is very simple.

With an origin chosen at A we can immediately write down the following expression for the B.M.:

$$M = Ax + B + Wz, \quad z = x - 2L. \tag{1.3.0}$$

The particular problem is described by $M_A = M_p$, $M_B = -M_p$, $M_C = M_p$ whence

$$M_A = M_p = B,$$
$$M_B = -M_p = 2AL + B, \tag{1.3.1}$$
$$M_C = M_p = 3AL + B + WL.$$

Hence $\qquad AL = -M_p, \quad W_c L = 3M_p.$

This latter relation gives the collapse load.

Exercise

Verify the following results (fig. 1.3.1):

$$M = Ax + B + (px^2/2), \quad p_c L^2 = 16Mp,$$
$$M = Ax + B, \quad W_c L = 8M_p.$$

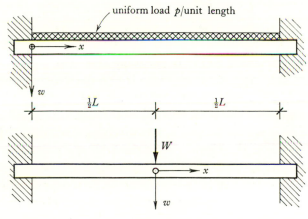

Fig. 1.3.1

Notes

1.2   The subscript $(\ )_c$ will be used without further comment to mean 'value of $(\ )$ at collapse'.

1.3   In general, given a beam, more than the required number of hinge sites exist and experience must be used to choose from amongst the possible hinge sites so as to give sufficient hinges for a mechanism and at the same time ensure that no other possible site is associated with a critical moment. A tool here is provided by the limit theorems which we discuss in 1.5.

1.4   The worked example above made no specific use of displacement information. This was because the presence of a mechanism with the chosen hinges was confirmed by inspection. In more complex problems, and especially with cases where it is intended to investigate by computer analysis, the presence of a mechanism cannot be confirmed so easily. One procedure is to write down an analogous set of displacement equations to the equilibrium equations. Then the condition for the existence of a mechanism with a chosen set of hinge locations is that the *rank* of the matrix of coefficients from these displacement equations should be *non-zero* (see (6.7) for a definition of rank).

1.4   An integral formula: rigid–plastic beams

Attention has already been drawn to the evident duality between w' and F on the one hand and w and M on the other. A further feature of this duality is brought out as follows. Consider the integral

$$\int_\alpha^\Omega \{w.(M)'' - M.(w'')\}\, dx, \tag{1.4.0}$$

where α, Ω are the end points of the beam.

Now, on integration by parts,

$$\int_\alpha^\Omega \{w.(M)'' - M.(w)''\}\, dx = w.M'|_\alpha^\Omega - M.w'|_\alpha^\Omega. \tag{1.4.1}$$

It is already evident and it is generally true that w and M can be expected to have discontinuities in their first derivatives. Despite this, the relation (1.4.1) continues to hold true. It is to be expected, too, that a step change in the value of the beam slope through a hinge will be noticed in w''. In fact it is easy to verify, intuitively at least, that if $[\![w']\!] = \theta$ then $\int (w)'' dx$ across the hinge $= \theta$. In the limit of a point concentration of a hinge, w'' will have a Dirac delta function ($\delta(x)$) behaviour of strength $\theta (= \int w'' dx)$ at a particular hinge site (see 0.6). So too for all hinges and so too for M'' at points of transverse load application, where there will be delta function behaviour of strength equal to the transverse load.

The interpretation of (1.4.1) is all important. First it should be pointed out that a differential equation, typically $L(u) = R$, has meaning only for points in the range α to Ω *with the end points excluded*. Then, with a careful regard as to signs, it can be seen that the terms linear in w in (1.4.1) collectively give the value of the external work done (force × displacement) by the loads and reactions as hinge rotations θ_i develop at collapse, whereas the terms linear in M together give the internal work done against the full plastic moments, M_p, at the respective hinges. These latter terms it should be noted are always additive since moment work is absorbed independently of the sign of the hinge rotation.

Here then is a useful alternative interpretation which can be placed on the rigid–plastic behaviour. The example treated in 1.3 can now be re-analysed thus: external work done $= 2W_cL\theta$, moment work absorbed $= M_p\{\theta + (\theta + 2\theta) + 2\theta\}$, where the hinge rotation at $A = \theta$.

Then equating
$$W_cL = 3M_p \tag{1.4.2}$$
as before.

Notes

1.5 The result (1.4.1) can alternatively be established by virtual work considerations. So too can some of the further results which we shall derive. But we shall avoid virtual work considerations and give a systematic development of our subject in terms of the differential equations and particularly shall exploit the very useful concept of discontinuous solutions.

1.6 By presenting the discussion from the discontinuous solution view point, the step to machine computation and matrix methods, which it is planned to discuss in a companion volume, will be quite natural. So too will be the inclusion within this same method of approach of the *two dimensional* problems of plate and shell structures.

1.7 The result (1.4.1) is a consequence of the fact that the operator $L(\ ) = (\ )''$ is self-adjoint (Ince (0.8)). All the equations which we shall study in this volume are self-adjoint. As a consequence various first integrals of these equations can be formed. These integrals are what we as engineers call *energy* integrals.

1.5 The limit theorems

A thorough-going treatment of rigid–plastic beams or frameworks traditionally depends upon a pair of theorems, the upper and lower bound, or limit, theorems. For our purposes we shall state the theorems, justify them by intuitive arguments rather than offer proofs and use them as the need arises. For *proofs* the reader should refer to the more specialized texts on rigid–plastic frameworks but especially Drucker (3.16).

There are three ingredients in the theorems. They are respectively considerations of *equilibrium, mechanism* and *yield*. For a collapse load to be the rigid–plastic collapse load there must be equilibrium between the external and internal forces at collapse, a mechanism must form with sufficient hinges for a deformation of the structure to occur and finally

2-2

yield must not be violated; in the case of the beam this latter is the statement that no moment *equals* (or *exceeds*) the local full plastic moment except at the chosen hinge sites, where the moment equals the full plastic moment. The limit theorems are concerned with the situation when only equilibrium and one of the two remaining conditions is known to be satisfied. The *upper bound theorem* states that the collapse load based on equilibrium and an assumed mechanism will be greater than or equal to the true collapse load. The *lower bound theorem* on the other hand states that the collapse load computed from the conditions that equilibrium and yield be satisfied will result in a collapse load which is less than or equal to the true collapse load. In each case the equality is true only when all three conditions of equilibrium, mechanism and yield are satisfied simultaneously.

Intuitively these results seem reasonable since in the case of an assumed mechanism, for this in fact to occur when it is not the true mechanism, some parts of the frame will require strengthening to prevent them collapsing before the assumed hinges form. Hence the collapse load will be over-estimated; that is, an unsafe upper bound on the collapse load will be obtained.

Conversely if the yield condition is known not to be violated then the load in equilibrium with this assumed system of internal forces (regarded as a collapse load) will be an underestimate, that is a safe lower bound on the collapse load, since it will in general imply fewer hinge sites than necessary for a mechanism.

All but the simplest types of rigid–plastic problems require application of the limit theorems to seek out the most likely mechanisms and help construct equilibrium force distributions. In practice the upper bound theorem is the more useful result because of the greater ease to assume a mechanism on the one hand as against constructing an equilibrium force distribution throughout the structure on the other.

In brief, what the limit theorems provide is a scheme for choosing from amongst the available partial solutions so as to ensure that the collapse load can be bounded from above and below by generally easily calculable values. The need for the limit theorems (or equivalent) becomes evident when we observe that the collapse load is the solution of a non-linear problem referred to the unknown hinge positions as parameters. The theorems suggest that the problem should not be looked at from this point of view but instead should be linearized by the methods proposed by the theorems.

In concluding this brief discussion of rigid–plastic beam behaviour we would urge the reader to work many problems on the subject, selected either from those at the end of the chapter or from the more specialized works given in the references (1.20).

1.6 Elastic and elastic–plastic beams

In the previous section the rigid–plastic moment/curvature relation was applied to beam *strength* problems. We now wish to generalize the discussion by incorporating the initial elastic response which characterizes the behaviour of all real materials and allows us in addition to derive information as to the *stiffness* as well as the strength of the structure.

The recent history of engineering structural calculations has been characterized by a strong current of development for load factor methods as opposed to safety factor methods. These terms require some defining. Load factor methods focus attention on the collapse load and seek to know its value for a given structure and load system in order to take some proportion of this load as the working load; then the ratio collapse to working load is defined as the load factor. Safety factor methods on the other hand aim to know the distribution of the forces in the structure, which is most often assumed to be in an entirely elastic condition, and then use a working load which will limit the actions at the worst stressed section to some suitable fraction of an assumed worst condition; for example, first attainment of plastic strains. The ratio worst : working loads is then termed the *safety factor*.

The load factor method then *global* is in the sense of requiring the whole frame to collapse whereas the safety factor method is *local*, requiring worst conditions at a critical section only. Inherently, the safety factor method applied to ductile structures is more conservative, requiring generally a heavier beam or frame for given loads than the load factor method. Very roughly, safety factor methods are elastically based where load factor methods are plastically based and they are for this reason often given quite separate treatment. But from various viewpoints and not least the computations themselves, there is much to be said for treating the two approaches side by side. This is the approach followed in the methods about to be discussed.

In the longer historical context, too, there are good reasons for treating the two approaches side by side. First collapse load studies were made by Galileo (1564–1642) and especially Mariotte (1620–84). The latter made observations on beams and plates at collapse and correctly established a number of results; for example, that a beam fixed at the ends and loaded with a central point load is twice as strong as the same beam with ends supported but unrestrained in rotation. These were basic strength studies. Mariotte too made a start on stiffness studies with the introduction of elasticity considerations. From Mariotte's time through to the present day there have been several waves of interest and development which have emphasized each of the load factor/strength approach and the safety factor/stiffness approach from time to time.

We now proceed to analyse elastic and elastic–plastic beam problems and to demonstrate the unity of our subject.

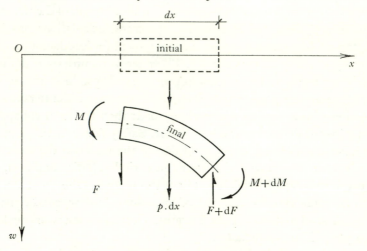

Fig. 1.7.0   Elastic/elastic–plastic beam element.

1.7   The fundamental equations: elastic and elastic–plastic beams

For the beam element shown in fig. 1.7.0 with exaggerated displacement and slope, the following equilibrium equations can be written down

$$M' = F, \quad F' = p. \tag{1.7.0}$$

Since the displacements are assumed to be small these equations are the same as for the rigid–plastic case.

The new information to describe the elastic material property is embodied in the proportionality between moment and curvature, which for small displacements takes the form

$$M = EI\frac{\mathrm{d}w^2}{\mathrm{d}x^2} \equiv \beta w'', \tag{1.7.1}$$

where E is Young's Modulus, I the section second moment of area and w the transverse displacement. If M and F are eliminated from (1.7.0–1) then

$$(\beta w'')'' = p, \tag{1.7.2}$$

or, if as is usually the case in practice, $\beta = \text{const.}$, then

$$\beta w^{\text{iv}} = p. \tag{1.7.3}$$

This governing differential equation is the basic equation of elastic and, as will be seen presently, of elastic–plastic beam behaviour, and is the fundamental equation of the present study.

1.8   The fundamental solution: elastic and elastic–plastic beams

The O.D.E. (1.7.3) can be integrated in quadratures to give

$$w = Ax^3 + Bx^2 + Cx + D + (px^4/4!\,\beta). \tag{1.8.0}$$

Here p and β have been assumed constant. The four arbitrary constants A, B, C and D reflect the fact that (1.7.3) is a fourth order O.D.E. The important point to note is that these four constants are, as yet, unknown and must be found by specifying *two* conditions at each end of the beam. It is never the case that these constants are found from, say three conditions at one end and only one at the other, or four and zero; always two and two. This is a reflexion of the fact that the equation (1.7.3) is of boundary value type. The terms $Ax^3 \ldots + D$ in (1.8.0) form the complementary function, the final term in p is a typical continuous particular integral.

Example

A uniform elastic beam, length L is fixed against rotation (encastré) and displacement at the two ends and loaded with a uniformly distributed load, total value W. Find the moment at the fixed end and the central displacement.

Now, generally,

$$w = Ax^3 + Bx^2 + Cx + D + \frac{Wx^4}{4!\,\beta L}, \tag{1.8.1}$$

and, with an origin at the end, the boundary conditions are $w = w' = 0$ at $x = 0$, L. The two conditions at $x = 0$ give $D = C = 0$. Then with $x = L$

$$\left. \begin{aligned} AL^3 + BL^2 + \frac{WL^3}{4!\,\beta} &= 0, \\[2mm] \text{and} \qquad 3AL^2 + 2BL + \frac{WL^2}{3!\,\beta} &= 0. \end{aligned} \right\} \tag{1.8.2}$$

Whence $$B = WL/24\beta,$$

and $$M_0 = \beta(w'')_0 = 2\beta B = \frac{WL}{12}. \qquad (1.8.3)$$

Then $$A = -\frac{W}{12\beta},$$

and $$w(L/2) = \frac{WL^3}{\beta}\left(-\tfrac{1}{12}\cdot\tfrac{1}{8}+\tfrac{1}{24}\cdot\tfrac{1}{4}+\tfrac{1}{24}\cdot\tfrac{1}{16}\right) = \frac{WL^3}{384\beta}. \qquad (1.8.4)$$

Exercise

If in the above example the end rotational restraint is removed, now termed a pin end or simple support, show that the central displacement is five times the previous value.

The above two examples are the simplest representatives of two classes of structure. The worked example is of an *indeterminate*, the exercise is of a *determinate*, beam. A distinction is drawn beween those structures where the forces (that is forces and moments) in the structure can be determined from statics alone (determinate structures) and those for which a knowledge of the material properties is required as well as statics (indeterminate structures). But in either case, if displacement information is required then the material properties must be made use of. Using the solution (1.8.0) does not require us to recognize whether the structure is of the one class or the other. But if the structure is determinate then the constants A and B can always be found independently of C and D (see note 1.10).

Notes

1.8   From (1.7.0, 1) we have $F = \beta w'''$ and $M = \beta w''$ and these relations will be used extensively. It will be assumed that $\beta = EI = $ const., for a particular member unless otherwise stated. F here is a typical shear force and M a typical bending moment.

1.9   The assumed directions for forces and displacements are as shown in fig. 1.7.0 and should be carefully noted.

1.10   The four constants A, B, C, D in the complementary function have physical meanings. As is easily verified, C and D are respectively the slope and displacement at the chosen origin while $2\beta B$ is the bending moment and $6\beta A$ the shear force at this same point. *This interpretation for the constants will be used repeatedly in what follows without further comment.*

1.9   Discontinuous solutions: force discontinuities; elastic beams

Thus far we have dealt only with single member problems. In extending the discussion for example to a beam continuous across an intermediate support we shall see that the concept of the discontinuous solution introduced for rigid–plastic problems finds a natural place here also. The reaction R at the support illustrated in fig. 1.9.0 can be described with precision as a discontinuity in the shear force at A, together with an associated zero displacement at A.

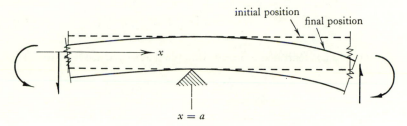

Fig. 1.9.0   Typical indeterminate support.

We must therefore inquire as to what the relevant (shear force) P.I. is for equation (1.7.1).

The argument proceeds exactly as in 0.3. For completeness we set out the computation in full below.

Now the physical interpretation in terms of discontinuities of the displacement at A is

$$\llbracket w \rrbracket = \llbracket w' \rrbracket = \llbracket w'' \rrbracket = 0, \quad \llbracket w''' \rrbracket = -R/\beta \quad \text{at} \quad x = a.$$

The minus arises from positive increments in F being downward and R is assumed directed upward.

Here the most general particular integral is

$$bx^3 + cx^2 + dx + e, \tag{1.9.0}$$

where the four constants b, c, d, e are to be found by satisfying the four discontinuity conditions listed above, at $x = a$.

Hence

$$\left.\begin{aligned}
ba^3 + ca^2 + da + e &= 0, \\
3ba^2 + 2ca + d &= 0, \\
6ba + 2c &= 0, \\
6b &= -R/\beta.
\end{aligned}\right\} \tag{1.9.1}$$

Solving for $b - e$,

$$b = -\frac{R}{6\beta}, \quad c = \frac{Ra}{2\beta}, \quad d = -\frac{Ra^2}{2\beta}, \quad e = \frac{Ra^3}{6\beta}, \tag{1.9.2}$$

whence D.P.I. for $[F] = -R$ is obtained by substituting (1.9.2) into (1.9.0), namely

$$-\frac{Rz^3}{3!\beta} \quad (z = x - a). \tag{1.9.3}$$

The more usual way of writing this result is for a gravity load W downward when

$$[F] = W \quad \text{implies a D.P.I. of} \quad Wz^3/3!\beta. \tag{1.9.4}$$

Exercise

Show that for a concentrated moment M applied at $x = a$, the required D.P.I. is $Mz^2/2!\beta$.

Example

Analyse a uniform fixed end single span beam, span L, with central point load W. Find the elastic moments and displacements.

With an origin at one end

$$w = Ax^3 + Bx^2 + \frac{Wz^3}{3!\beta} \quad \left(z = x - \frac{L}{2}\right).$$

The fixity conditions at $x = 0$ have already been satisfied by $C = D = 0$ and finally $w = w' = 0$ at $x = L$ give

$$\left. \begin{array}{l} AL^3 + BL^2 + \dfrac{WL^3}{48\beta} = 0, \\[2mm] 3AL^2 + 2BL + \dfrac{WL^2}{8\beta} = 0, \end{array} \right\} \tag{1.9.5}$$

whence

$$A = -\frac{W}{12\beta}, \quad B = \frac{WL}{16\beta},$$

and

$$M_0 = 2\beta B = \frac{WL}{8}, \quad w\left(\frac{L}{2}\right) = \frac{AL^3}{8} + \frac{BL^2}{4} = \frac{WL^3}{48\beta}. \tag{1.9.6}$$

Note

1.11   There is a symmetry in this example which has not been exploited. It is left to the reader to rework the example using the range of integration from 0 to $L/2$, with an origin at the midpoint. But beware of the argument that the displacement is an even function – it is not true (see 1.15).

Example

Consider now a uniform elastic beam of length $2L$, fixed at the ends and passing over an unyielding support at mid-span fig. 1.9.1. The right span is loaded with a mid-span load, W. Draw the bending moment diagram.

With an origin at the left end then

$$w = Ax^3 + Bx^2 - \frac{Rz_1^3}{3!\beta} + \frac{Wz_2^3}{3!\beta}, \tag{1.9.7}$$

where $z_1 = x - L$, $z_2 = x - 3L/2$ and R is the unknown support reaction.

There are three unknowns in (1.9.7), A, B, R. The three conditions to be satisfied are zero central support displacement, and zero displacement and slope at the right end. Namely

$$\left.\begin{aligned} AL^3 + BL^2 &= 0, \\ 8AL^3 + 4BL^2 - \frac{RL^3}{6\beta} + \frac{WL^3}{48\beta} &= 0, \\ 12AL^2 + 4BL - \frac{RL^2}{2\beta} + \frac{WL^2}{8\beta} &= 0. \end{aligned}\right\} \tag{1.9.8}$$

Whence
$$64\beta A = W, \quad 64\beta B = -WL, \quad 2R = W. \tag{1.9.9}$$

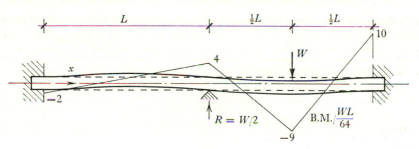

Fig. 1.9.1   B.M. drawn on tension side of beam.

Then the B.M. at salient points can be found as

$$M_0 = 2\beta B = -\frac{WL}{32}, \quad M_L = \beta(6AL + 2B) = \frac{WL}{16},$$

$$M_{2L} = \beta(12AL + 2B) = \frac{5WL}{32}.$$

Hence the B.M. diagram (fig. 1.9.1).

Exercise

Investigate the D.P.I. for a uniformly distributed load beginning at $x = a$. This is the beam equivalent of the problem investigated in 0.3 and can be obtained from the result obtained there for the limiting case $P \to 0$. The result is, for a distributed load p, a D.P.I. given by $pz^4/4!\beta$. The reader may verify the result from first principles.

1.10   Discontinuous solutions: displacement discontinuities; elastic–plastic beams

Thus far our discussion of discontinuities has been limited to force discontinuities, typically the point transverse load. There is an important physical occurrence of a *displacement-type discontinuity* and this is the sudden change in slope of a member through a plastic hinge.

When a ductile beam is subjected to increasing bending moment it is observed that at a definite value of the B.M. the maximum moment capacity of the section is reached and further rotation occurs at this constant maximum moment, the full plastic moment, M_p. The subsequent examination of the member reveals a very localized zone where practically all the rotation has occurred and this is idealized to a point discontinuity in the slope of the member.

The conditions at the hinge are, clearly, w' discontinuous by some (unknown) amount and the shear force, bending moment, displacement continuous and $|\text{B.M.}| = M_p$. Now by an exactly similar argument to that used in 0.3 we can investigate suitable D.P.I.s which will ensure the correct sort of discontinuous behaviour.

The result is that a slope discontinuity of θ at $x = a$ implies a D.P.I. of

$$\theta \cdot z \quad (z = x - a). \qquad (1.10.0)$$

The plastic hinge concept is central to discussion of elastic–plastic and rigid–plastic theory of structures. These theories, as has already been said, are collapse or ultimate load theories and lead on to significantly different design philosophies compared with elastic theory. From the point of view of the analyses, the only superficial difference between the two theories, elastic and plastic, with the present methods, is the occurrence of θz terms associated with plastic hinges in the equations describing the latter.

Collapse in the plastic theory usually means that sufficient hinges have formed in the structure to cause it to become a mechanism and incapable of sustaining any further increase in load. But collapse may also be deemed to have occurred if the displacements become too great. The customary treatment of the plastic theory assumes a rigid–plastic response with the result that information about displacements is lacking. If this information is required a separate elastic–plastic study must be made.

In the present development of the theory of structures, all three possible approaches of rigid– or elastic–plastic or purely elastic response produce a computational scheme which retains the maximum of common working and this clearly is a considerable asset. Let us illustrate this final remark with an example embracing all three schemes.

Example

Trace the load/displacement history for the beam considered in 1.3 but now regarded as an elastic–plastic beam. This is a fixed end beam, span $3l$, loaded with point load W at l from right-hand support. Choose as an origin the left support. Then, with $l = L$:

Stage I, elastic

The deflected shape is given by

$$w = Ax^3 + Bx^2 + (Wz^3/3!\,\beta) \quad (z = x - 2l). \tag{1.10.1}$$

There are two remaining unknowns to be obtained from the right-hand end conditions of $w = w' = 0$ at $x = 3l$, namely

$$\begin{bmatrix} 27 & 9 \\ 27 & 6 \end{bmatrix} \begin{bmatrix} Al \\ B \end{bmatrix} = \frac{Wl}{8} \begin{bmatrix} \frac{1}{6} \\ \frac{1}{2} \end{bmatrix}. \tag{1.10.2}$$

Hence
$$162A\beta = -7W, \quad 9B\beta = Wl, \tag{1.10.3}$$
and

$$M_A = M_0 = 2\beta B = \frac{2Wl}{9}, \quad M_B = M_{2l} = -\frac{8Wl}{27}, \quad M_C = M_{3l} = \frac{4Wl}{9}, \tag{1.10.4}$$

whence
$$M_A < M_B < M_C. \tag{1.10.5}$$

Stage II, elastic–plastic

The first hinge therefore forms at C when $W = W_1$, and

$$4W_1 l = 9M_p \quad \text{or} \quad W_1 l = 2 \cdot 25 M_p. \tag{1.10.6}$$

The displacement at B is then

$$\Delta_1 = 8Al^3 + 4Bl^2 = \tfrac{2}{9} Ml^2/\beta. \tag{1.10.7}$$

With further increase in load a hinge rotation develops at C. Hence the second equation in (1.10.2) no longer applies, being the condition for zero hinge rotation. Instead W increases until $W = W_2$ and $M_B = -M_p$ when the second hinge forms, at B, and

$$12Al\beta + 2B\beta = -M_p. \qquad (1.10.8)$$

The first of (1.10.2), (1.10.8) and $M_C = M_p$, that is,

$$18Al\beta + 2B\beta + W_2 l = M_p, \qquad (1.10.9)$$

then give

$$28B\beta = 11M_p, \quad 28W_2 l = 81M_p, \quad \text{or} \quad W_2 l = 2 \cdot 89 M_p. \qquad (1.10.10)$$

The associated displacement at B, Δ_2, is given by

$$\Delta_2 = \frac{8}{21} \frac{M_p l^2}{\beta} = 0 \cdot 381 \frac{M_p l^2}{\beta}. \qquad (1.10.11)$$

As a check, $M_A = 2B\beta = \frac{11}{14}M_p$, confirming that a hinge forms at B before that at A.

Thereafter further increase in W will cause a hinge rotation to develop at B until, as the moment at A reaches the full plastic value, the beam will collapse.

For this final phase of the loading cycle a hinge discontinuity must be included in the deflected shape.

Hence

$$w = Ax^3 + Bx^2 + \frac{Wz^3}{3!\beta} + \theta \cdot z \quad (z = x - 2l), \qquad (1.10.12)$$

where θ is the unknown hinge rotation at B.

The first equation of (1.10.2) should now have a θl term added to it, the equations (1.10.8–9) remain unchanged except that now $W = W_3$, but additionally, the moment at A has now reached the fully plastic value and a rotation is on the point of developing at A.

Hence $w'_A = 0$ as ensured by (1.10.12) and

$$2\beta B = M_p. \qquad (1.10.13)$$

Then with (1.10.2), as modified, (1.10.8, 9, 13) we have

$$2\beta B = M_p; \quad W_3 l = 2M_p; \quad \Delta_3 = \frac{2}{3} \frac{M_p l^2}{\beta}, \qquad (1.10.14)$$

and the entire load/displacement history for the load point B can be plotted as in fig. 1.10.0.

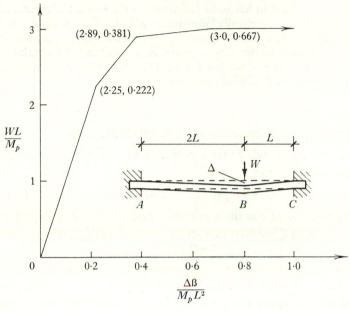

Fig. 1.10.0

Exercise

(1) The simple single span fixed end beam with central point load W is loaded to collapse. Show that in the elastic range the end and midspan moments are of equal magnitude and find the collapse load and central displacement at collapse.

Ans. $$W_c L = 8M_p, \qquad 24\Delta_c \beta = M_p l^2.$$

(2) The same beam but loaded with a u.d. load p/unit length over the full span. Show that the collapse load is given by $p_c L^2 = 16M_p$, and $\Delta_c = $ the central displacement at collapse by $12\Delta_c \beta = M_p L^2$.

Notes

1.12 *Formulation of the problem.* In all cases, the solution of a given problem begins with the writing down of an expression for the displacement in terms of a complementary function added to which is a list of terms arising from the evident discontinuities which can be seen to be present. It is important to note however that excluded from this list are any discontinuities, for example a plastic hinge, which might be present at either end of the beam. The 'beam' in this context need not be a single span beam, it could equally well be multi-span, but the point to note is that the *end points and their discontinuities are excluded* for the good reason that they are already automatically included in the arbitrary constants of the complementary function.

1.13 The choice of origin has some influence on the process of solution but only marginally so and generally any origin will do. Occasionally what is required for later use from the solution may influence the choice; for example, immediate knowledge of the displacement at a point can be obtained by placing the origin there and solving for D. The origin placed elsewhere will require a little further calculation to obtain the result.

1.11 Summary of discontinuous particular integrals

All the particular integrals so far introduced to deal with discontinuous happenings are of the form

$$Xz^n/n! \quad (z = x - a), \qquad (1.11.0)$$

$x - a$ being the site of the discontinuity. Specifically, if $[\![\text{slope}]\!] = \theta$, then $X = \theta, n = 1$; $[\![\text{bending moment}]\!] = M, X = M/\beta, n = 2$; $[\![\text{shear force}]\!] = F, X = F/\beta, n = 3$; $[\![\text{distributed loading}]\!] = p, X = p/\beta, n = 4$; and so on in a sequence. Although we have not yet encountered it, there is a useful P.I. for discontinuity in *displacement* (Δ) for which $X = \Delta, n = 0$.

In what follows, appropriate choices from the above list will be made to deal with the various physical situations which arise.

For emphasis, too, we repeat that any expression which is a function of z is omitted for $z < 0$.

1.12 Further discontinuous solutions: simultaneous discontinuities

The occurrence of simultaneous discontinuities can best be seen through examples. Consider the two span elastic beam shown in fig. 1.12.0.

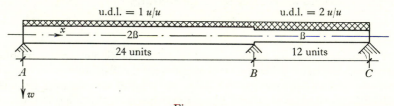

Fig. 1.12.0

We shall first write down the expression for the displaced form and then explain the terms

$$w = ax^3 + cx + \frac{x^4}{4!\,2\beta} + \frac{pz^4}{4!\,\beta} + \frac{Fz^3}{3!\,\beta} + \frac{Mz^2}{2!\,\beta}. \qquad (1.12.0)$$

1.12 *Simultaneous discontinuities*

Now to explain the significance of each term. The first two terms are the surviving pieces of the C.F. once the conditions at A have been satisfied. The third term is the P.I. for the u.d.l. in span AB; the p term, with p unknown, will be made to take care of the change in u.d.l.s from AB to BC. Finally the shear force and bending moment-like terms in F and M, F and M being unknowns, can be made to reflect the presence of a support at B and continuous BM through B. Hence there are three discontinuity type terms arising at B and this is the circumstance referred to as simultaneous discontinuities.

In the absence of a support at B, both the moment and the shear at B must be continuous. But these conditions require discontinuity in w'' and w''' since $\beta w''$ and $\beta w'''$ are continuous but β is discontinuous at B. The presence of the support means an actual discontinuity in F due to the reaction R, but this can conveniently be lumped in with the S.F.-like term in F already included in (1.12.0). Once the form (1.12.0) has been scrutinized to ensure that all the physical effects are included, the equation formulation and solution can proceed.

In the present case there are five unknowns a, c, p, F, M and five conditions: $w_B = 0$, equation satisfied across B, $[\![M_B]\!] = 0$, $w_c = w_c'' = 0$.
In turn these conditions yield

$$24^2 . a + c = -12 \times 24, \tag{1.12.1}$$

$$(\tfrac{1}{2} + p) = 2, \quad p = 1.5, \tag{1.12.2}$$

$$\left.\begin{array}{c} 2\beta(6 \times 24 \times a) + \dfrac{24^2}{2} = \beta\left(6 \times 24 \times a + \dfrac{24^2}{4\beta}\right) + M, \\[2ex] \beta(36^3 a + 36c) + \dfrac{36^4}{48} + 3 \times \dfrac{12^4}{48} + \dfrac{12^2 M}{2} + \dfrac{12^3}{6} \times F = 0, \end{array}\right\} \tag{1.12.3}$$

$$\beta(6 \times 36 \times a) + \dfrac{36^2}{4} + 3 \times \dfrac{12^2}{4} + M + 12F = 0, \tag{1.12.4}$$

whence $\qquad \beta a = -\tfrac{13}{16}, \quad \beta c = 180, \quad M = 27, \quad F = -\tfrac{189}{8}.$

Hence $\qquad\qquad M_B = 2\left(6 \times 24 \times a\beta + \dfrac{24^2}{4}\right) = 54u,$

and hence the B.M. diagram.

With the addition of θz terms to (1.12.0) and appropriate choice for hinge positions, the elastic–plastic behaviour of the beam fig. 1.12.0 can also be investigated.

Note

1.14 The recognition of *the occurrence of simultaneous discontinuities* and their exploitation is the clue to the whole further development of the subject. The topic will be taken up again in the discussion of frameworks. But clearly multi-span beams can be treated in just the same fashion as in the previous example.

1.13   Discontinuous solutions: two further elastic examples

The aim in the present treatment of structures is to develop a comprehensive method of solution, based on discontinuous solutions, so as to bring all structural types within scope. Experience may indicate, or time may prove, the methods to be more suitable for a specific range of problems. For example, the properties of single members may be needed to start a moment distribution or provide information for a computer program. The following two examples set out to derive information required for a subsequent application to moment distribution. Even within this limited sphere of application, the discontinuity methods are at least as satisfactory as any alternative approach.

Example: Stiffness and carryover factors I

Consider the member shown in fig. 1.13.0 with an external moment M_a applied as shown.

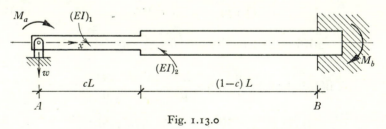

Fig. 1.13.0

If such a member forms part of an elastic framework a knowledge of M_a/θ_a and M_b/M_a, the stiffness and carry-over factor respectively, will be required if a moment distribution, see (1.20.4), is to be attempted.

For a step change in the value of member inertia, proceeding as in the example in 1.12, and ensuring S.F. and B.M. continuity at the step change in I, then

$$w = Ax^2 + Bx^2 + Cx + D - \frac{[I]}{I_2}A(x-cl)^3 - \frac{[I]}{I_2}(3Acl+B)(x-cl)^2.$$

$$\text{(1.13.0)}$$

Here $[I] = I_2 - I_1$.

1.13 *Discontinuous solutions: elastic example*

For the particular beam configuration of fig. 1.13.0, with $I_2 = 2I_1$, $c = 0.5$, (1.13.0) reduces to

$$w = Ax^3 - \frac{M_a}{2EI_1}x^2 + Cx - \frac{A}{2}\left(x - \frac{l}{2}\right)^3 - \tfrac{1}{4}(3Al + 2B)\left(x - \frac{l}{2}\right)^2. \quad (1.13.1)$$

This expression contains two unknowns A, C and it has still to satisfy the conditions at b, i.e. $w = w' = 0$.

Using these two conditions gives

$$\left.\begin{array}{l} \tfrac{4}{3}Al^3 + \tfrac{7}{8}Bl^2 + Cl = 0, \\[4pt] \tfrac{3}{4}Al^2 + \tfrac{3}{5}Bl + \tfrac{2}{5}C = 0. \end{array}\right\} \qquad (1.13.2)$$

In (1.13.2) $B = -M_a/2EI_1$ and finally

$$9Al = -5B, \quad 24C = -11Bl. \qquad (1.13.3)$$

Hence the stiffness,

$$\frac{M_a}{\theta_a} = -\frac{2BEI_1}{C} = \frac{48}{11}\frac{EI_1}{l}, \qquad (1.13.4)$$

and the carryover factor,

$$\frac{M_b}{M_a} = -\frac{2}{3}\frac{BEI_2}{-2BEI_1} = \frac{2}{3}. \qquad (1.13.5)$$

Example: stiffness and carryover factors II

As a further example of practical importance consider the beam fig. 1.13.1. Then, for elastic conditions, the parameters as defined in fig. 1.13.1, and as yet general boundary conditions,

$$w = Ax^3 + Bx^2 + Cx + D + \frac{[I]}{I_1}(x - cl)^2\{A(x - cl) + (3Acl + B)\}$$
$$+ \frac{F(x - (1 - c)l)^3}{3!\,EI_2} + \frac{M(x - (1 - c)l)^2}{2!\,EI_2},$$

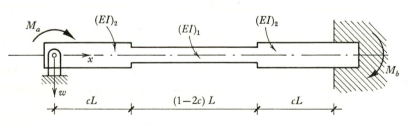

Fig. 1.13.1

35

where
$$F = -6EA[I]\left(1 + \frac{[I]}{I_1}\right), \tag{1.13.6}$$

and
$$M = -(6A(1-c)l + 2B)E[I]\left(1 + \frac{[I]}{I_1}\right). \tag{1.13.7}$$

As before, $[I] = I_2 - I_1$.

The equations are written in the above manner in order to emphasize their structure.

Consider now a particular case, $I_2 = 2I_1$, $c = 0.25$, then with $w = D = 0$ at a, and $M_a = -2BEI_2$ and known, the two boundary conditions at b give

$$\left.\begin{array}{l} 4.5Al^2 + 4Bl + \frac{8}{3}C = 0, \\ 4.5Al^2 + 3Bl + C = 0, \end{array}\right\} \tag{1.13.8}$$

whence
$$5C = -3Bl, \quad 15Al = -8B, \tag{1.13.9}$$

and stiffness
$$\left.\begin{array}{l} = \dfrac{M_a}{\theta_a} = -\dfrac{2BEI_2}{-\frac{3}{5}Bl} = \dfrac{20}{3}\dfrac{EI_1}{l}, \\[3mm] = \dfrac{M_b}{M_a} = \dfrac{8}{5} - 1 = 0.6. \end{array}\right\} \tag{1.13.10}$$

carryover factor

Clearly, (1.13.1) and (1.13.6) can be very easily programmed for a digital computer should occasion demand.

These two examples illustrate how efficiently single member problems can be dealt with. Again, although these examples have been of elastic situations, they could equally well have been for plastic, either rigid- or elastic- members.

1.14   Continuous variations of section: member properties

If the EI is continuously varying then it may become more economical to revert to the fundamental differential equation (1.7.2), namely

$$(EIw'')'' = 0, \tag{1.14.0}$$

rather than approximate the continuous variation with a series of step changes.

For example if the beam is rectangular and of linearly varying depth then $EI \propto x^3$. If $EI = kx^3$ then (1.14.0) can be integrated in quadratures to give

$$w = \frac{1}{k}\left(-a\ln x + \frac{b}{2x}\right) + cx + d. \tag{1.14.1}$$

1.14 *Continuous variation of section*

If the beam is as depicted in fig. 1.14.0 and $I_b = (3)^3 I_a$, namely the beam is of constant width and beam depth at b three times that at a, then, with the choice of origin as indicated in fig. 1.14.0 it follows that $k = 8EI_a/L^3$ and the ends of the beam are $x = L/2, 3L/2$. Suppose knowledge of the stiffness for end a is required. Then the requirements are $w_a = 0$, $w_a'' = -M_a/EI_a$; $w_b = w_b'' = 0$, and M_a is the applied clockwise test moment at A.

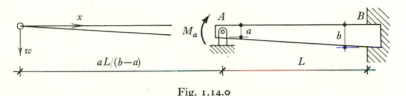

Fig. 1.14.0

Then

$$
\left.
\begin{aligned}
&\frac{L^3}{8EI_a}\left(-a\ln\frac{L}{2}+\frac{b}{L}\right)+\frac{cL}{2}+d = 0, \\[2mm]
&\frac{aL}{2}+b = -M_a, \\[2mm]
&\frac{L^3}{8EI_a}\left(-a\ln\frac{3L}{2}+\frac{b}{3L}\right)+\frac{3cL}{2}+d = 0, \\[2mm]
&\frac{L^3}{8EI_a}\left(-\frac{2a}{3L}-\frac{2b}{9L^2}\right)+c = 0.
\end{aligned}
\right\}
\qquad (1.14.2)
$$

Solving (1.14.2) we obtain

$$aL = 2 \cdot 058 M_a,$$
$$b = -2 \cdot 029 M_a,$$
$$c = 0 \cdot 114 M_a,$$

and the stiffness at a

$$= -M_a/\theta_a,$$
$$= 9 \cdot 52(EI_a/L).$$

In this and similar cases a, d do not have their usual physical meanings. Other types of section variation can be investigated by the same approach.

1.15 Discontinuous solutions: a bridge problem; elastic behaviour

The structure shown in fig. 1.15.0 is a simplified version of a bridge structure such as might be used at a split level trunk road junction. It consists essentially of a stiff deck supported on comparatively slender

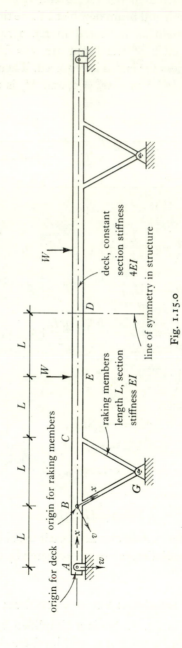

Fig. 1.15.0

origin for deck

A
B

origin for raking members

x
v
w

x

C
E
D

W
W

L L L L L L

raking members
length L, section
stiffness EI

G

deck, constant
section stiffness
$4EI$

line of symmetry in structure

raked columns. The most important part of the structure is the central span. Initially we shall investigate the symmetrical structure under a symmetrical load system, and hence need consider only one half of the structure.

The difference in principle between this problem and preceding ones is the need for two coordinate origins in this case to describe the structure. Then with the choice of origins as indicated on fig. 1.15.0.

For *ABCED* the transverse displacement w is given by

$$w = Ax^3 + Cx + \frac{F_B z_1^3}{3!\,\beta} + \frac{M_B z_1^2}{2!\,\beta} + \frac{F_C z_2^3}{3!\,\beta} + \frac{M_C z_2^2}{2!\,\beta} + \frac{W z_3^3}{3!\,\beta}. \qquad (1.15.0)$$

For *BGC* the transverse displacement v is given by

$$v = ax^3 + bx^2 + cx + \frac{F_G z_1^3}{3!\,EI}, \qquad (1.15.1)$$

where
$$z_i = x - iL, \quad \beta = 4EI.$$

These displacement descriptions already embody the conditions at A and the zero displacement condition for v_B. The unknowns A, C, F_B, M_B, F_c, M_c; a, b, c, F_g are ten in number and are to be found from the following ten conditions:

(1) B is a fixed point in space $\qquad w_B = 0$

(2) Slope continuity at B between ABC $\qquad w'_B = v'_B$
     and BG

(3) Moment continuity at B between $\qquad \beta[w''_B] = EIv''_B$
     ABC and BG

(4) C is a fixed point in space $\qquad w_c = 0$

(5) C is a fixed point in space $\qquad v_c = 0$ $\qquad\qquad$ (1.15.2)

(6) Slope continuity at C $\qquad w'_c = v'_c$

(7) Moment continuity at C $\qquad \beta[w''_c] = EIv''_c$

(8) Slope zero at D (symmetry condition) $\qquad w'_D = 0$

(9) Shear force zero at D (symmetry $\qquad w'''_D = 0$
     condition)

(10) G is a fixed point $\qquad v_g = 0.$

If the unknowns and equations are written down in the order thus enumerated then, in matrix form, the equations are

$$\mathbf{AS} = \mathbf{L}, \qquad (1.15.3)$$

where

$$\mathbf{S} = \left[AEI, \frac{CEI}{L^2}, F_B, \frac{M_B}{L}, F_C, \frac{M_c}{L}, aEI, \frac{bEI}{L}, \frac{cEI}{L^2}, F_g \right]^T,$$

$$\mathbf{L} = W[0, 0, 0, 0, 0, 0, 0, -0{\cdot}125, -0{\cdot}25, 0]^T,$$

and

$$\mathbf{A} = \begin{bmatrix} 1 & 1 & 0 & 0 & 0 & 0 & 0 & 0 & 0 & 0 \\ 3 & 1 & 0 & 0 & 0 & 0 & 0 & 0 & 1 & 0 \\ 0 & 0 & 0 & 1 & 0 & 0 & 0 & 2 & 0 & 0 \\ 8 & 2 & \frac{1}{24} & \frac{1}{8} & 0 & 0 & 0 & 0 & 0 & 0 \\ 0 & 0 & 0 & 0 & 0 & 0 & 8 & 4 & 2 & \frac{1}{6} \\ 12 & 1 & \frac{1}{8} & \frac{1}{4} & 0 & 0 & -12 & -4 & -1 & -\frac{1}{2} \\ 0 & 0 & 0 & 0 & 0 & -1 & 12 & 2 & 0 & 1 \\ 48 & 1 & 1{\cdot}125 & 0{\cdot}75 & 0{\cdot}5 & 0{\cdot}5 & 0 & 0 & 0 & 0 \\ 6 & 0 & 0{\cdot}25 & 0 & 0{\cdot}25 & 0 & 0 & 0 & 0 & 0 \\ 0 & 0 & 0 & 0 & 0 & 0 & 1 & 1 & 1 & 0 \end{bmatrix}.$$

Solving the system (1.15.3) we obtain the solution vector

$$W(-0{\cdot}004528, 0{\cdot}004528, 0{\cdot}804346, -0{\cdot}050724, -1{\cdot}695652,$$

$$0{\cdot}137681, -0{\cdot}016304, 0{\cdot}025362, -0{\cdot}009057, 0{\cdot}282608)^T. \quad (1.15.4)$$

From (1.15.4) it is now a simple matter to construct the S.F. and B.M. diagrams and the deflected shape.

For example, the midspan B.M. is given by

$$M_D = 6A\beta(4L) + 3F_B L + M_B + 2F_c L + M_c + WL$$

$$= (96(-0{\cdot}004528) + 3(0{\cdot}804346) - 0{\cdot}050724$$

$$+ 2(-1{\cdot}695652) + 0{\cdot}137681 + 1)\, WL$$

$$= -0{\cdot}326WL. \quad (1.15.5)$$

Also, the moment at C in CD, $(M_c)_D$, is given by

$$(M_c)_D = 6A\beta(2L) + F_B L + M_B + M_c$$

$$= (48(-0{\cdot}004528) + (0{\cdot}804346) - 0{\cdot}050724 + 0{\cdot}137681)\, WL$$

$$= 0{\cdot}674WL. \quad (1.15.6)$$

As a check, $(M_c)_D + |M_D| = WL$, which should be the case since this is the free span moment for the central span of $4L$ and the loading shown.

1.15   *Discontinuous solutions: bridge problem*

The central displacement w_D is given by

$$w_D = 64AL^3 + 4CL + 1{\cdot}125\frac{F_B L^3}{EI} + 1{\cdot}125\frac{M_B L^2}{EI}$$
$$+ \frac{1}{3}\frac{F_c L^3}{EI} + 0{\cdot}5\frac{M_c L^2}{EI} + \frac{WL^3}{24EI}$$
$$= 0{\cdot}1214\frac{WL^3}{EI}. \qquad (1.15.7)$$

Treated as an encastré beam the central $4L$ span under the given loading produces $0{\cdot}25\,WL, 0{\cdot}75\,WL, WL^3/12EI$ corresponding to $(1.15.5, 6, 7)$ respectively. Hence, due to the flexibility of the supporting structure $ABCG$, the maximum moment has been reduced by approximately 10 % but at the cost that the maximum displacement has been increased by some 50 %.

Exercises

(1) For the same structure and loading as in fig. 1.15.0 except that the load on the right half is reversed in sign to produce the antisymmetric problem, show that the solution vector is given by

$$W(-0{\cdot}001882,\, 0{\cdot}001882,\, 0{\cdot}334336,\, -0{\cdot}021084,\, -0{\cdot}929216,$$
$$0{\cdot}057228,\, -0{\cdot}006777,\, 0{\cdot}010542,\, -0{\cdot}003765,\, 0{\cdot}117469)^T.$$

Superposition of this result with $(1.15.4)$ allows unsymmetrical load systems to be studied.

Hint   If conditions 8 and 9 are exchanged for $w_D = w_D'' = 0$, the result follows.

(2) Now consider the structure of fig. 1.15.0 to be rigidly fixed at D and consider the left half only. Show that the solution vector is given by

$$W(-0{\cdot}001157,\, 0{\cdot}001157,\, 0{\cdot}205555,\, -0{\cdot}012962,\, -0{\cdot}619444,$$
$$0{\cdot}035185,\, -0{\cdot}004166,\, 0{\cdot}006481,\, -0{\cdot}002314,\, 0{\cdot}072222)^T$$

and hence that $M_D = 0{\cdot}2894WL, F_D = 0{\cdot}5585W$.

Hint   If condition 9 is exchanged for $w_D = 0$, the result follows.

Note

1.15   If different ratios of inertias between the deck and supports, or different spans, are required with the same structural layout, it is a simple matter to make the changes to the coefficients in the matrix **A** in $(1.15.3)$. To make actual changes to the layout of the type envisaged in the Exercise 2 is also a simple matter of change of one equation, the second last to read $w_D = 0$ rather than $w_D''' = 0$.

41

1.16 Symmetry and evenness

Many structures in practice possess symmetries; the structure in fig. 1.15.0 for example. In treating these problems, the symmetries can be immediately used to solve for or eliminate some unknowns. But some care must be exercised in not equating symmetry in a problem with evenness in the functions of the independent variable which appear. In discussing discontinuities in the manner we have done, we have endowed our independent length coordinate, x, with a certain 'time-like' quality since we have adopted the convention $f(z) \equiv 0$ if $z < 0$. As a result, symmetry and evenness are not equivalent and in fact evenness will play no part in the further development.

Readers can best satisfy themselves of the correctness of this latter statement by examining the consequences of assuming the equivalence of symmetry and evenness in some simple examples.

1.17 Conclusion

Only the simplest problems of plastic and elastic beams have been discussed thus far. More complicated problems can be handled in exactly similar fashion however. So too can practical problems involving, for example, support flexibility.

A *beam* in the present context is an assembly of one or more members joined to form a composite *straight* member. In the next chapter more general jointing of members will be discussed, thus introducing the subject of *frameworks*. There is some further discussion of beam problems in chapters 4 and 5, where other governing equations are developed, but as regards allocation of pages to topics we have consciously devoted quite a proportion of the space available to discussing beams with the $\beta w^{\mathrm{IV}} = p$ governing equation since this is the problem most often met with in practical situations. Considerable facility should be developed in handling these problems and acquiring a feel for the type of result to be expected. An important feature in this connexion is checks on answers obtained. It is recommended that the procedures here developed be used exclusively in analysing a problem. Then the simple overall statical checks remain as satisfying and easily applied independent checks on results.

1.18 Further exercises

Establish the following results (1)–(4). The subscript $(..)_c$ refers to rigid plastic collapse. All other values are for elastic conditions.

(1)

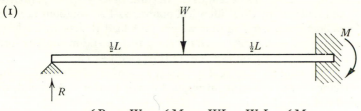

$$16R = 5W, \quad 16M = 3WL, \quad W_cL = 6M_p.$$

(2)

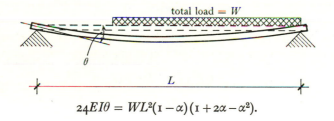

$$6\theta EI = W\alpha L^2(1-\alpha)(2-\alpha), \quad 48EI\delta = W\alpha L^2(3-4\alpha^2), \quad W_cL\alpha(1-\alpha) = M_p.$$

(3)

total load $= W$

$$24EI\theta = WL^2(1-\alpha)(1+2\alpha-\alpha^2).$$

(4)

$$M_A L^2 = Wab^2, \quad M_B L^2 = Wa^2b, \quad 6EI\delta_c = b^2M_p(b \geqslant a), \quad W_c ab = 2M_p L.$$

43

(5) Prove the three moment theorem: $M_{n-1} + 4M_n + M_{n+1} = 0$, where M_i is the hogging moment at support n (that is producing tension in the top of the beam over the support) for a long uniform equal multi-span elastic beam and with the spans under consideration not subject to any external load.

(6) A uniform elastic beam of length $2l$, flexural rigidity EI, is supported at the ends and midspan by three identical pontoons. The pontoons are each of area $\mathscr{A}$ and float in sea water of unit weight γ. A load W acts on the beam a distance cl from one end. If $\mathscr{A}\gamma = 3EI/l^3$ show that the beam support force provided by the middle pontoon is given by $W(c(3-c^2)+2)/8$.

(7) In the notation of fig. 1.13.0, if $c = 0.25$, $I_2 = 9I_1$ find the stiffness and carry over factor for the test moment applied to the smaller section end, the larger section end being built in.

Ans. Stiffness $= 6EI_1/l$, carry over factor $= 1$.

1.19   Solutions and comments on exercises

The exercises (1)–(4) are simple applications of the foregoing principles.
   Typically Ex. (2)

$$w = Ax^2 + Cx + \frac{Wz^3}{3!\beta} \quad (z = x - \alpha L).$$

Then
$$AL^3 + CL + \frac{W(1-\alpha)^3 L^3}{6\beta} = 0,$$

$$6AL + \frac{W(1-\alpha)L}{\beta} = 0.$$

From which
$$6C\beta = WL^2(1-\alpha)(-1+2\alpha-\alpha^2+1)$$
$$= WL^2\alpha(1-\alpha)(2-\alpha),$$

and so on.

(5) Consider the spans $(n-1)/n$ and $n/(n+1)$, let the reaction at support $n = R$ and the spans be of length a. With an origin at support $n-1$ then

$$w = Ax^3 + Bx^2 + Cx + D + R(x-a)^3/6\beta.$$

At $n-1$
$$D = 0, \quad M_{n-1} = 2\beta B.$$

At n
$$Aa^3 + Ba^2 + Ca = 0,$$
$$(6Aa + 2B)\beta = M_n.$$

At $n+1$
$$8Aa^3 + 4Ba^2 + 2Ca + (Ra^3/6\beta) = 0,$$
$$(12Aa + 2B)\beta + Ra = M_{n+1}.$$

Eliminate A, C, R when $M_{n-1} + 4M_n + M_{n+1} = 0$.

   Kármán and Biot (1.20) give an application of this rather famous equation, due to Clapeyron, with use of the finite difference calculus.

It is of interest too that Clebsch (1.20) considering this same multi-span beam problem was the first to use a discontinuous approach in dealing with a structural problem. In English language texts the name of Macaulay (1919) (1.20) is more often associated.

(6) Denote the pontoon reactions by R_1, R_2, R_3.

$$\text{Then} \qquad w = Ax^3 + Bx^2 + Cx + D + \frac{W(x - cl)^3}{6\beta} - \frac{R_2(x - l)^3}{6\beta}.$$

Now
$$B = 0, \quad D = -2Al^3, \quad R_2 = 12\beta A + (2 - c)\,W,$$

and
$$19\beta A = W[-c^3 + 11c - 14], \quad \text{whence} \quad R_2 = W(c(3 - c^2) + 2)/8.$$

The second result derives from the displacement/reaction relation at the origin end pontoon, namely

$$\mathscr{A}\gamma w_0 = R_1 = -\beta w_0'',$$

or
$$\mathscr{A}\gamma D = -6\beta A \quad \text{and so on.}$$

See Timoshenko (1.20).

(7) The general expression quoted in (1.13.0) can be used directly, although a first principles approach should be attempted if (1.13.0) has not already been verified.

Extensive charts for stiffness and carry over factors for a variety of beams with variable section properties are given by Cross and Morgan (2.26), Lightfoot (1.20) and Gere (1.20).

1.20 References

There is no book which develops the subject in the manner of the present chapter 1. The following books are suggested for supplementary reading and alternative approaches.

Baker, Sir John and Heyman, J. *Plastic Design of Frames*, vol. 1, Cambridge University Press, London (1969). A modern text dealing with rigid plastic frameworks. Problems with answers. S.I. units throughout.

Clebsch, A. *Theorie der Elasticität Fester Körper*, Teubner (1862). Here is probably the first explicit use of the discontinuity methods in structural mechanics. The problem treated is the beam on n supports, on p. 390.

Crandall, S. H. and Dahl, N. G. (editors). *An Introduction to Mechanics of Solids*, McGraw-Hill, New York (1959). This modern American text, at a first course level, makes some use of discontinuity methods and attributes the approach to Macaulay.

Gere, S. M. *Moment Distribution*, Van Nostrand, New York (1963). Many examples with answers, but only elastic problems treated.

Kármán, T. v. and Biot, M. A. *Mathematical Methods in Engineering*, McGraw-Hill, New York (1940). Many interesting structural examples although the primary interest is in the mathematics. Problems with hints and answers.

Lightfoot, E. *Moment Distribution*, Spon, London (1961). Similar in depth of treatment to Gere (3) but deals with a wider range of problems. More slanted toward machine computation.

Macaulay, W. H. 'A note on the deflection of beams.' *Messenger of Mathematics* (1919), **48**, 129. Macaulay was presumably unaware of Clebsch's treatment, see above, of the same type of problem and by a similar method. This paper introduced discontinuity ideas to English speaking readers although in a very brief and incomplete manner.

Mariotte, E. *Traité du mouvement des eaux*, J. Jombert, Paris (1700). This diminutive volume contains a detailed account of Mariotte's experiments on a whole range of topics, including the strength of beams. The relevant portion is Part v, Discours ii.

Timoshenko, S. P. *Strength of Materials*, vol. i, Van Nostrand, New York (1940), 2nd edition. This book, and the vol. ii, are a mine of detailed results covering a wide range of problems in small displacement elasticity. Main approach used is energy methods. Problems for the reader.

Timoshenko, S. P. *History of Strength of Materials*, McGraw-Hill, New York (1953). A compact, well-balanced history of our subject.

Timoshenko, S. P. and Young, D. H. *Theory of Structures*, McGraw-Hill, New York (1965), 2nd edition. Very detailed within the limited scope of treating only elastic problems. Problems for the reader.

2

PLANE FRAMEWORKS

2.0 Introduction

Thus far only single or composite *straight* members have been considered. Attention will now be given to *frameworks* formed by joining members together in other configurations.

Frequently in the literature separate developments of frame theory are given for each of the usually assumed material properties. Thus some treatments deal only with elastic frameworks, others with rigid–plastic only. This is an undesirable separation since there are many features common to elastic (generally working load) and rigid–plastic (ultimate load) conditions for a framework and understanding can be increased by noting common features and where the behaviour and methods of solution differ. In addition, few actual structures of any size in practice are considered solely from the point of view of a single material property idealization.

The present treatment of plane frameworks aims to discuss all such theories in a connected fashion beginning with the simplest, the rigid–plastic theory, then moving into elastic and finally elastic–plastic considerations. This is the same sequence as was considered for beams.

Precisely the same fundamental differential equations and solutions which were developed for the beam apply to the framework. Thus the equilibrium and material property equations are unchanged. But in addition, *changes of direction* of a member through a *joint* must now be studied and the consequences for the integration procedure evaluated.

Thus the only new feature of the description of a framework as compared with a beam structure arises from changes of direction as we move the reference point from one member to the next through a joint. This is because the differential equations do not contain any reference to axial components of force which, although usually present, are not of sufficient magnitude as to affect the behaviour of the member. Later we shall study the case when the axial force does affect the behaviour when a further phenomenon is possible, instability or buckling, in addition to any strength or stiffness considerations.

In the present linear theory then, the G.D.E. is, as it were, blind to the presence of axial forces. The frame, however, is influenced by them since, for example, the axial force in a vertical member is transmitted to a horizontal joining beam as shear force and this is an action which influences the frame behaviour. Information about the axial forces themselves can only be obtained by consideration of the overall equilibrium of portions or the whole of the framework, a *global* property. This is in contrast to the usual conditions met with so far which are point properties, i.e. *local* properties. Our object will be to so arrange the writing down of these overall equilibrium, or sway equations as we shall frequently call them, that the axial forces, being unknowns, are excluded.

For the most part we shall be concerned with loading applied to our plane frameworks which causes the distortion of the framework to take place within the plane of the framework. Typically, a vertical plane frame subject to gravity forces will satisfy these conditions. Further, there are good practical reasons why we should, if at all possible, seek to have the frame loaded in this manner since it is generally then strongest.

2.1 Topology of the structure: statical indeterminacy and kinematic freedom

Thus far only beam structures consisting of straight members joined to form a composite straight member have been considered. Frameworks are far and away a more interesting class of structure. A framework is any system of interconnected members; for simplicity we shall restrict the discussion to *plane* frameworks, that is frameworks in which all the members lie in one plane. Usually horizontal and vertical members account for the bulk of the members. The horizontal members will be termed *beams*, the vertical members, *columns*. A member which is inclined to the vertical can be assigned to either category although if it is near vertical it is convenient always to refer to such members as columns. The main concern, too, is with rigid jointed structures.

A ring in the present context is a closed loop traced out through the members of the framework and the supporting structure. The minimum number of rings (R) required to describe a structure, that is ensure that every member is included in at least one ring, is a *topological invariant* for the structure and is characteristic of the member interconnexions. Closely related to this minimum number of rings is the *statical indeterminacy* (I) of the framework. This is the excess of the minimum number of force *unknowns*, over and above those obtainable from statics, which is

48

just capable of defining the forces and moments at every cross-section in the structure. More precisely, if there are Q known components of internal force or moment in the whole structure then

$$I = 3R - Q. \qquad (2.1.0)$$

An example of a known internal component is a zero moment at a (real) hinge or the plastic moment associated with a known plastic hinge.

If $I = 0$, then the structure is said to be *determinate*; if $I > 0$, then redundant (or indeterminate); if $I < 0$ then *deficient* (or a mechanism). To be precise, these conditions are necessary but not sufficient conditions since $I = 0$, for example, does not rule out part of the structure being deficient and part redundant. Further tests must be applied to eliminate such a possibility.

When a framework is determinate the internal force system can be established by statics alone. The stress–strain relations for the material need not be used to find the internal *force* system although clearly they must be used to obtain any displacement information. Expressed another way, if the structure is determinate the equilibrium and force–displacement equations describing the framework become decoupled. The number of force unknowns equals the number of equilibrium equations and can be solved for independently of the force–displacement equations. If the structure is indeterminate the two sets of equations must be solved simultaneously.

Although the degree of statical indeterminacy is an invariant for a given structure and is independent of the loading system (except in so far that symmetry might give certain of the redundants to be zero), the choice of which forces are to be regarded as redundants is to some extent arbitrary. The only requirement is that if all the chosen redundants are set to be zero the remaining structure must be statically determinate. More precisely, the resulting structure must be just stiff rather than redundant in some parts and as a consequence, a mechanism in other parts. Herein lies the difficulty in defining the logic for a suitable choice.

The energy theorems in applied statics use the redundants as the basic unknowns. This choice gives a minimum number of unknowns in many situations. The present method of analysis uses more than this (minimum) number of unknowns.

In addition to statical considerations there are also kinematic (displacement) considerations which should be discussed at this time. Throughout the present treatment of frameworks individual members will be thought of as inextensible; they can bend but are assumed not to stretch or compress. This is a justifiable working approximation. As a consequence, if two non-collinear members are each jointed to different fixed points and then to a common point to form a very simple structure, this common point can be seen to be a *fixed* point because of member inextensibility. Such a frame is said to have no kinematic freedoms. If instead of being connected one to the other they are jointed to opposite ends of a third member then the structure acquires a single kinematic freedom. Thought of as bars freely hinged together, the system is a mechanism with a single degree of freedom. If a further member is added to the chain of members, a further freedom is added. This number of *kinematic freedoms* is clearly closely related to the topology of the structure.

The above example is of a single ring structure. A structure with a multi-ring topology presents some difficulties in formulating a general expression for kinematic freedoms. But in most of the framed structures in bending met with in practice the number of kinematic freedoms (K) is given by

$$K = 2J - M, \tag{2.1.1}$$

where J is the number of joints other than support points and M is the actual number of members, no fictitious members being supplied by the supports.

A member here joins two joints. If a continuous column has beams spanning into it then the column length between adjacent beams constitutes a member and the continuous column length in all will comprise a number of members.

The importance of recognizing the number of kinematic freedoms lies in the need, in general, to formulate as many equations as there are kinematic freedoms in the framework under study. These equations are obtained from the conditions of equilibrium of entire members or parts of the framework – they are global equations. The remaining, and generally more numerous, equations are of two types. First there are the familiar local conditions of continuity and discontinuity, a zero support displacement or occurrence of a plastic moment at a hinge site for example, such as occur in beam problems. Secondly the conditions of inextensibility on the displacements of joints connected by a member. The equations arising from kinematic and inextensibility considerations are peculiar to frameworks and can best be illustrated with examples. This we shall do a little later.

As a final point we should note that in some situations, notably under symmetrical loading conditions, symmetrical frames which possess kinematic freedoms may not have these freedoms brought into play (excited) by the loading. It is then as if the frame did not possess these freedoms. Alternatively, it is clear that if the kinematic equations were formulated, they would be automatically satisfied.

Part I   Simple frameworks
2.2   Simple frameworks I: rigid–plastic

The rigid–plastic theory of frameworks as developed here is the simplest theory of frameworks and is the extension of the theory already developed for beams to other configurations of members. It is an ultimate load theory and is a remarkably accurate theory for predicting the collapse load of bare ductile frameworks up to perhaps a few storeys in height. The theory has nothing to say about the stiffness of the structure, that is the displacements produced, merely the strength.

It is assumed that all members of the frame are not subject to any form of instability and that the fully plastic moment is the only material property of interest. Any cladding of a real framework is ignored in the present theory although in so far as this might increase the full plastic moment, its effects could be taken account of.

To repeat what has already been said for beams, according to the rigid–plastic theory, the members and joints, and hence the frame, remain rigid and undistorted until the bending moment at sufficient points of the structure reaches the full plastic value for a (stiff) mechanism to be formed. Any further increase in load can then not be withstood.

In principle, finite hinge rotations can then develop; in practice our interest is to examine the frame just as the bending moment at the final hinge to form reaches the full plastic value but before any significant displacements have developed. The meaning of 'any significant displacements' is that the displacements are assumed small enough for the equilibrium equations to be written in terms of the initial coordinates, with sufficient accuracy.

2.3   The fundamental equations: rigid–plastic

As for the beam element (fig. 1.10), the fundamental equilibrium equations are

$$M' = F, \quad F' = p, \tag{2.3.0}$$

or $$M'' = p, \tag{2.3.1}$$

together with the yield criterion

$$|M| \leqslant M_p. \tag{2.3.2}$$

The displacements enter through the material property of rigidity; thus

$$w'' = 0. \tag{2.3.3}$$

As earlier, F is the S.F., M the B.M. and w the transverse displacement.

2.4 The fundamental solutions: rigid–plastic

The differential equation of equilibrium (2.3.1) can be integrated in quadratures to give

$$M = Ax + B + \text{P.I.s.} \tag{2.4.0}$$

Likewise (2.3.3) integrates to give

$$w = ax + b + \text{P.I.s.} \tag{2.4.1}$$

A very useful restriction is to require that all external loads are applied as point transverse loads. The effect of this restriction is to force all hinge sites to be points of shear force discontinuity (for example, a joint or a point of load application). If in practice a distributed load is encountered, at least initially this can be replaced by a suitable statically equivalent series of point loads.

In a manner exactly as for beams, discontinuous particular integrals can be introduced. Thus a discontinuity in the shear force, the fundamental discontinuity, for a step change in the value of F at $x = a$, $z = x - a$, necessitates addition of a particular integral Fz to (2.4.0), together with the usual convention that the term is to be neglected for $z < 0$.

2.5 A rigid–plastic example: coordinate systems

To take full advantage of discontinuous solution techniques it is very often essential to use as few coordinate systems to describe the dependent quantities, such as M and w, as possible. For instance, the single choice of point 1 in fig. 2.5.0 as coordinate origin will allow the coordinate axis to be 'bent' at joints 2 and 4, and used for the entire frame, provided that it is noticed that what is axial thrust in 1–2 becomes shear in 2–4 and vice versa. Namely, the act of *bending* the coordinate axis in this way will produce a discontinuity in the shear force at 2, but will leave the bending

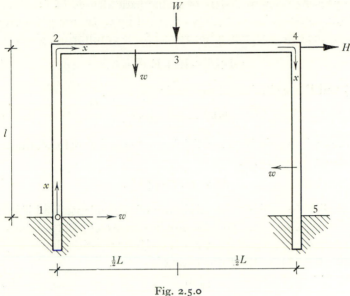

Fig. 2.5.0

moment continuous, as desired. The same may be true of displacements but discussion of this point is delayed until elastic frames are considered.

Suppose the discontinuities in the S.F. at 2 and 4 to be F_2 and F_4 respectively. Then

$$M = Ax + B + F_2 z_2 + W z_3 + F_4 z_4 \qquad (2.5.0)$$

is the most general description of the bending moment in the frame. This is true independently of any rigid–plastic considerations. To satisfy the tenets of the rigid–plastic theory, sufficient full plastic moments must develop to produce a mechanism (or mode of deformation). In the present case four such hinges (three if a symmetrical mode) are required and there are five possible hinge sites, being the points 1–5 inclusive.

The correct choice of hinges is that choice which minimizes the collapse load (see limit theorems, 1.5). Suppose the choice of hinge sites at points 1, 3, 4, 5 is made.

Then

$$\left. \begin{aligned}
M_1 &= B = M_p, \\
M_3 &= A(l + \tfrac{1}{2}L) + M_p + \tfrac{1}{2}F_2 L = -M_p, \\
M_4 &= A(l + L) + M_p + F_2 L + \tfrac{1}{2}WL = M_p, \\
M_5 &= A(2l + L) + M_p + F_2(L + l) + W(\tfrac{1}{2}L + l) + F_4 l = -M_p.
\end{aligned} \right\} \qquad (2.5.1)$$

The present framework has one kinematic degree of freedom. The associated equation is most easily obtained by considering the beam cut free from the columns, when horizontal force equilibrium gives

$$2A + F_2 + W + F_4 = -H. \tag{2.5.2}$$

If F_2 and F_4 are eliminated we obtain

$$\tfrac{1}{2}WL + Hl = 6M_p. \tag{2.5.3}$$

Once the loading ratio W/H and the frame shape L/l are fixed, the collapse load follows from (2.5.3).

Also

$$Al = \tfrac{1}{2}WL - 4M_p. \tag{2.5.4}$$

We shall have attained the minimum of the collapse load if it can be ensured that at no other possible hinge sites except the assumed hinge sites does the bending moment reach the fully plastic value.

In the present case the requirement is

$$|M_2| \leqslant M_p. \tag{2.5.5}$$

Now
$$\begin{aligned} M_2 &= Al + B \\ &= \tfrac{1}{2}WL - 3M_p. \end{aligned} \tag{2.5.6}$$

Whence the limits implied by (2.5.5) are

$$\tfrac{1}{2}WL - 3M_p \geqslant -M_p, \quad \tfrac{1}{2}WL - 3M_p \leqslant M_p$$

or
$$4M_p \geqslant \tfrac{1}{2}WL \geqslant 2M_p. \tag{2.5.7}$$

If (2.5.3) is used in (2.5.7) we obtain

$$\tfrac{1}{4} \leqslant Hl/WL \leqslant 1. \tag{2.5.8}$$

Provided this inequality is satisfied then the assumed mechanism is the actual one. If $Hl/WL < \tfrac{1}{4}$ then the mechanism can be seen to be hinges at 1, 2, 3 and 4. If $Hl/WL > 1$ hinges will form at 1, 2, 4 and 5.

As with the beam, so too for the framework, the duality between M and w can be exploited in an integral relation to give a work equation for finding the collapse load condition once a mechanism is decided upon. Virtual work considerations can equally well be used to establish such a relation. This method is well established and extensively discussed in the literature. The references cited at the end of the chapter should be consulted for details. But work methods of themselves do not give a systematic

2.5 Rigid–plastic example: coordinate systems

method for describing the bending moment distribution throughout the frame, information which the present method does provide and which is urgently needed if the computations are to be made on a digital computer.

Notes

2.1 It has not been necessary to discuss the displaced form in the above problem. The reason is that inspection of the problem has reassured us that the chosen system of hinges does in fact produce a mechanism. In a more complicated case, and especially where computer analysis is being attempted, a check using the solution of $w'' = 0$ may conveniently be used to confirm that a mechanism will form with an assumed set of hinges.

2.2 Let us again observe that the basically non-linear problem has here been very conveniently linearized by assuming a mechanism.

2.6 Simple frameworks II: elastic

An elastic framework is a framework in which the value of the bending moment at all points is less than M_y, the yield moment. All deformation under load is fully recoverable on removal of the load; the frame as it were has a perfect memory. Most frameworks in the working load range remain fully elastic, indeed this may be a design requirement.

2.7 The fundamental equations: elastic

As for the elastic beams (1.7) the member describing equations are

$$M' = F, \quad F' = p, \quad \beta w'' = M, \qquad (2.7.0)$$

or

$$(\beta w'')'' = p, \qquad (2.7.1)$$

where

$$\beta = EI.$$

2.8 The fundamental solutions: discontinuous particular integrals

The fundamental equations are integrable in quadratures and in the case of $\beta = $ const. we obtain

$$w = Ax^2 + Bx^2 + Cx + D + \text{P.I.} \qquad (2.8.0)$$

The four arbitrary constants A, B, C, D have the same physical meanings as in the beam case, namely, apart from a scale factor, they are the shear force, bending moment, slope and displacement at the origin, respectively.

The discontinuous particular integrals, too, for the framework are precisely the same as for the beam. However the frequency of use and the combinations of particular integrals are more numerous in frameworks than in beams.

Let us remind the reader that all the discontinuous particular integrals for physical variables are of the form $Xz^n/n!$ where for shear force $[F]$, $X = F/\text{ß}$, $n = 3$; bending moment $[M]$, $X = M/\text{ß}$, $n = 2$; slope $[\theta]$, $X = \theta$, $n = 1$; displacement $[\Delta]$, $X = \Delta$, $n = 0$. For $n > 3$, distributed uniform load $[p]$ gives $X = p/\text{ß}$, $n = 4$, and so on. Here, as earlier, $\text{ß} = EI$, E is Young's modulus and I the second moment of area for the member section.

Note

2.3   There is a temptation to define the solution (2.8.0) as

$$w = \sum_{i=0}^{3} \frac{c_i x^i}{i!},$$

instead of as we have done without the factorial. On balance we feel that the presentation is simpler without the factorial and have for this reason omitted it.

2.9   Simultaneous discontinuities: joints: elastic frames

Beams, either single or multi-span, are straight and a single coordinate origin clearly can suffice for the whole member. When a number of members are connected to form a framework, it is usual to introduce a new coordinate origin for each member. In many cases this is unnecessary. Where two members are jointed together, typically at right angles to one another, then, as has already been described in 2.4, the x coordinate line can be 'bent' round from one member and used for the adjacent member, but at the cost in an elastic or elastic–plastic framework of introducing possibly three additional unknowns. These unknowns represent quantities which are discontinuous as the corner is passed through.

Consider the specific case of a rigid beam–column junction, the beam horizontal and the column vertical and with common beam and column sections. Suppose a coordinate origin has been chosen at the foot of the column. Then this same origin can be used for the beam provided that the following two discontinuous particular integrals are added to the expression for the frame displaced form. First a term $R(x-h)^3/3!\,\text{ß}$, secondly a term $\Delta(x-h)^\circ$.

2.9 *Simultaneous discontinuities*

Here the dimension h is the column height and it will be seen that the terms indicate a discontinuous shear force and displacement. The reason is simply that although the slope and bending moment across the rigid joint are continuous, the shear force and displacement are not.

In fact, if the frame has no kinematic freedoms, such as happens in many symmetrical frames symmetrically loaded, then $\Delta \equiv 0$, but the shear force discontinuity term must always be included. The third possible discontinuity is one of bending moment *type* in cases where members of unequal section are jointed together, or where more than two members are jointed together. The processes used can best be appreciated with an example.

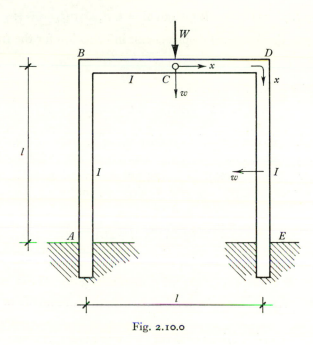

Fig. 2.10.0

2.10   Elastic frames with no joint displacements: I simple portal

Consider the simple portal shown in fig. 2.10.0. If the coordinate system indicated is adopted then

$$w = Ax^3 + Bx^2 + D + \frac{R_D z_D^3}{3! \, \text{ß}}, \tag{2.10.0}$$

with $\text{ß} = EI$ and $z_D = x - l/2$.

The description (2.10.0) already incorporates the slope zero condition

57

at C, namely $C = 0$. There remain four unknowns and they are found from the conditions

$$\text{shear at } C = \frac{W}{2}, \quad ßw_c''' = \frac{W}{2} = 6Aß,$$

$$w_D = 0, \quad \frac{Al^3}{8} + \frac{Bl^2}{4} + D = 0,$$

$$w_E = 0, \quad \tfrac{27}{8}Al^3 + \frac{9Bl^2}{4} + D + \frac{R_D l^3}{6ß} = 0,$$

$$w_E' = 0, \quad \tfrac{27}{4}Al^2 + 3Bl + \frac{R_D l^2}{2ß} = 0. \tag{2.10.1}$$

Solving we have

$$12Aß = W, \quad 12Bß = -Wl, \quad 96Dß = Wl^3, \quad 16R_D = -W, \tag{2.10.2}$$

and hence all the force and displacement information for the frame.

Exercise

Show that for the frame fig. 2.10.0

$$-6M_c = Wl, \quad 12M_D = Wl, \quad -24M_E = Wl.$$

If the beam in the previous exercise had a section different from the column, say $(ß)_{beam} = n \times (ß)_{column}$, then a term $Mz^2/2ß_c$ must be added to the description (2.10.0), M being an unknown, and the additional equation comes from the moment continuity condition at D, namely

$$[\![ßw'']\!]_D = 0$$

or $$M = (n-1)(3Al+2B). \tag{2.10.3}$$

2.11 Elastic frames with no joint displacements: II the mill frame

Consider the frame shown in fig. 2.11.0, with the column of non-uniform section.

Symmetry considerations confirm that there is no tendency for either of joints 3 or 5 to displace. Hence we are in a position to write down the expression (2.11.0) for the displaced form of the half frame 1-2-3-4 with an origin chosen at 1. Now

$$w = Ax^3 + Bx^2 + \frac{Rz_2^3}{3!\,ß} + \frac{Mz_2^2}{2!\,ß} + \frac{3 \cdot z_3^4}{4!\,ß} + \frac{F \cdot z_3^3}{3!\,ß} \tag{2.11.0}$$

with $ß = EI$ and $z_2 = x - 30$, $z_3 = x - 40$.

The five unknowns A, B, R, M and F have the following significance. A and B are the remaining portions of the C.F. after fixity at 1 is ensured;

2.11 *Elastic frames, no joint displacements*

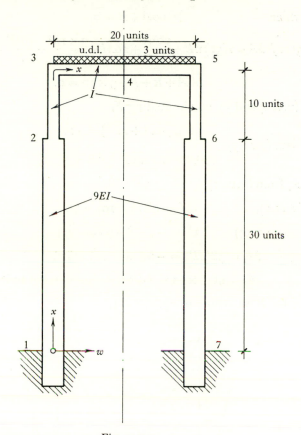

Fig. 2.11.0

R, M are unknowns of force and moment dimensions and will allow the junction 2 to be described; F is the shear force discontinuity at 3. The second last term is the distributed beam loading term.

The five unknowns are found from the following five conditions. Physical continuity of shear and moment at 2 (2 equations); zero displacement at 3 (1 condition); the symmetry conditions of zero slope and shear at 4 (2 conditions).

Proceeding now to write each of these equations in turn:

$$9ß(6A) = ß(6A) + R, \quad R = 48Aß, \tag{2.11.1}$$

$$9ß(6A \times 30 + 2B) = ß(6 \times A \times 30 + 2B) + M, \quad M = 16ß(90A + B), \tag{2.11.2}$$

$$40^3ßA + 40^2ßB + \frac{10^3R}{6} + \frac{10^2M}{2} = 0,$$

59

from which $\qquad 60A + B = 0,$ $\qquad\qquad$ (2.11.3)

$$3(50)^2\,\text{ß}A + 2(50)\,\text{ß}B + \frac{20^2R}{2} + 20M + \frac{3\times 10^3}{6} + \frac{10^2F}{2} = 0, \quad (2.11.4)$$

$$6\text{ß}A + R + F + 3\times 10 = 0. \qquad\qquad (2.11.5)$$

From (2.11.4 and 5), on eliminating F,

$$\text{ß}(4320A + 42B) = 100 \qquad\qquad (2.11.6)$$

when with (2.11.3)

$$18\text{ß}A = 1 \quad \text{and} \quad 3\text{ß}B = -10, \qquad (2.11.7)$$

whence, from (2.11.1) $\qquad 3R = 8,$

from (2.11.2), $\qquad 3M = 16(15 + 10) = 400,$

and from (2.11.5) $\qquad 3F = -1 - 8 - 90 = -99.$

Finally, $\qquad M_1 = 9\text{ß}(2B) = -60$ moment units,

$$M_2 = 9\text{ß}(6\times 30\times A + 2B) = 30 \text{ moment units,}$$

whence the B.M. diagram, fig. 2. 11. 1.

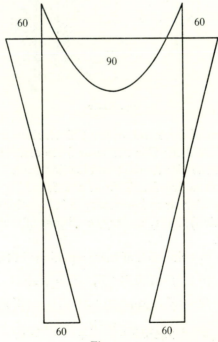

60   60

90

60   60

Fig. 2.11.1

2.11 Elastic frames, no joint displacements

The change in the dimension $2\text{–}6 = 2(A(30)^3 + B(30)^2) = -3000/ß$ length units, which is a spreading apart of the points 2 and 6. This information would be needed in the design of the frame if, for example, a moving gantry was spanning from 2–6.

Notes

2.4   The above example is typical of the calculation for a simple framework. The problem is first described by a coordinate system and studied for symmetries – which may reveal important information such as zero joint displacements. Then the description of the displaced form is written down which contains as unknowns all the evident discontinuities and load terms, along with the C.F. Finally the conditions which particularize the given problem can be written down and the unknowns solved for.

Clearly the most important single step in the process is writing down the displaced form. Some experience is needed here but this can best be gained by solving simple problems to begin with and at all stages checking the number of unknowns against the number of conditions which must be imposed.

2.5   The following notation will shortly be used: $(\ldots)_+$ will indicate the value of the bracketed quantity just past the particular point being considered and $(\ldots)_-$ for the value just before. The implication is that the quantity is discontinuous.

2.12   Frames with kinematic freedoms I

If the frame or the loading is not symmetrical then joints do displace. In this case there is a need to write down further equations, equal in number to the degrees of kinematic freedom.

Consider the square fixed base elastic portal of uniform section shown in fig. 2.12.0. This frame possesses one kinematic freedom, a side sway displacement, and, although the framework has an axis of symmetry, the loading does not and hence joint displacements will develop. We shall obtain a solution by considering the half frame ABC since we infer that under the loading shown, point C is a point of zero transverse displacement and bending moment.

Now the general expression for the displaced form in the present case is

$$w = Ax^3 + Bx^2 + (Rz^3/3!\,ß) + \Delta(z)^0. \qquad (2.12.0)$$

Note the absence of a term in H, this is because the force H is applied to a *joint*. The fixity conditions at A have already been introduced. There remain four unknowns A, B, R and Δ.

The antisymmetry which exists in the loading allows us to deduce that G is a point of contraflexure (zero moment) and zero displacement (w) and tells us also the shear at A is $-H/2$. Hence

$$6ßA = -H/2. \qquad (2.12.1)$$

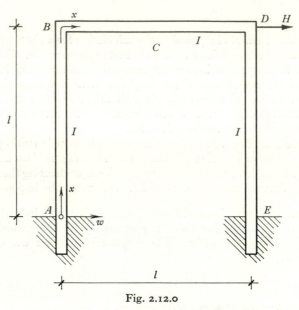

Fig. 2.12.0

We require three further conditions and they are $w_{B+} = 0$, namely the inextensibility condition, meaning that column AB is incompressible to our degree of approximation and $w_c = w_c'' = 0$. Hence

$$Bl^2 + \Delta = \frac{Hl^3}{12ß},$$

$$\tfrac{9}{4}Bl^2 + \frac{Rl^3}{48ß} + \Delta = \frac{9}{32}\frac{Hl^3}{ß}, \qquad \Bigg\} \qquad (2.12.2)$$

$$2B + \frac{Rl}{2ß} = \frac{3}{4}\frac{Hl}{ß}.$$

Finally, $\qquad 7ßB = Hl \quad$ and $\quad M_A = 2ßB = \tfrac{2}{7}Hl. \qquad (2.12.3)$

$$M_B = (6Al + 2B)ß = -\tfrac{3}{14}Hl, \quad w_B' = \frac{Hl^2}{28ß}$$

and the sidesway displacement

$$-\Delta = \frac{5}{84ß}Hl^3. \qquad (2.12.4)$$

2.12 *Frames with kinematic freedoms I*

Here we are dealing with elastic frames and hence all effects are additive; superposition of effects is permissible. Later we shall use the results of 2.10 and 2.12 in combination to discuss the elastic–plastic behaviour of the frame considered in 2.5 for rigid–plastic behaviour.

Exercise

Repeat the above calculations for the case span $= 2l$ and pin ground supports and show that

$$M_B = \frac{Hl}{2}, \quad w_B' = \frac{Hl^2}{6\beta}, \quad w_{B-}'' = \frac{Hl^3}{3\beta}.$$

2.13 Frames with kinematic freedoms II: overall equilibrium equations

Let us now consider a framework which has no symmetries. The solution in this case makes use of all the features required by quite general frameworks and we shall give a comprehensive discussion of the solution for this reason. The particular frame chosen has uniform section throughout, is fixed to one foundation and pinned to the other.

With the choice of origin at A as shown in fig. 2.13.0 we can immediately write down the following expression for the displaced form

$$w = Ax^3 + Bx^2 + \frac{R_B z_B^3}{3!\beta} + \Delta_B + \frac{9 \times z_E^3}{3!\beta} + \frac{R_c z_c^3}{3!\beta} + \Delta_c. \quad (2.13.0)$$

The terms appearing in (2.13.0) have the following significance: A, B complementary function; R_B, Δ_B the pair of expected discontinuities at B, the z_E term arises from the 9^u load and R_c, Δ_c are the expected discontinuities in shear and displacement at C. Note that the 2^u load does not appear explicitly in (2.13.0) because it is applied to a joint.

There are six unknowns in (2.13.0). The six conditions to be satisfied are as follows. First, two conditions describing the pin at D, $w_D = w_D'' = 0$. Then three equations describing the inextensibility of the three members in the frame and finally one equation describing the equilibrium of member BC when cut free from the frame. This latter equation is associated with the single degree of kinematic freedom possessed by the frame.

Before discussing these equations in detail let us discuss *inextensibility* requirements in general. These arise from our working assumption that only bending effects contribute to deformation.

Generally speaking, a joint in a framework will have *two* components of displacement, unless there is a member framing into this joint whose

63

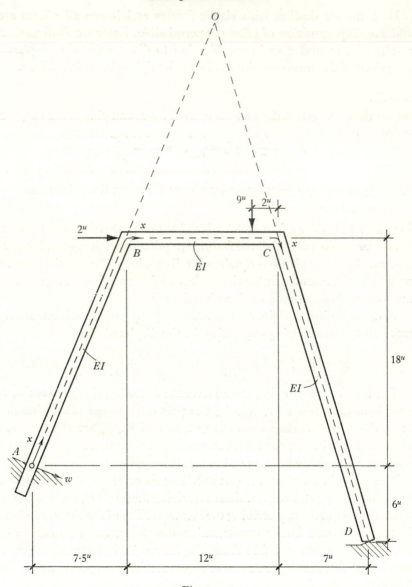

Fig. 2.13.0

other end is a fixed point when there will be only *one* component of displacement. But the only displacement we manipulate directly through equations and unknowns is the member *transverse* displacement. There appears to be some information lacking and this is the *axial* displacement

64

of a given member. However, by the nature of the problem, the axial movements are expressible in terms of the transverse displacement once the topology of the framework is known, through the *inextensibility* requirements.

Consider a typical joint, i, in a framework as shown in fig. 2.13.1. The coordinate directions are indicated. Quantities relating to the member

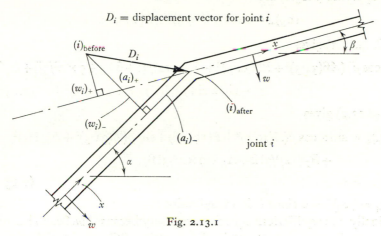

Fig. 2.13.1

before i will be denoted with a $(\ldots)_-$, those after by $(\ldots)_+$. The joint is shown in its displaced position, having moved a vector distance **D** under the action of the loads. The transverse component of displacement is denoted by the usual $(w_i)_-$, the axial component by $(a_i)_-$. Remembering the positive direction for w, we obtain as the resolution of **D**

$$(w_i)_+ = (w_i)_- \cdot \cos(\alpha - \beta) - (a_i)_- \cdot \sin(\alpha - \beta), \qquad (2.13.1)$$

and

$$(a_i)_+ = (a_i)_- \cdot \cos(\alpha - \beta) + (w_i)_- \cdot \sin(\alpha - \beta). \qquad (2.13.2)$$

In moving to the joint $i+1$, the axial component at $(a_i)_+$ remains unchanged and becomes $(a_{i+1})_-$ because of inextensibility and in this way the axial component of displacement of each member can be expressed in terms of the component for the preceding, and ultimately in terms of a zero axial component member.

Let us now apply these expressions to the framework fig. 2.13.0. From the geometry it will be found that $AB = 19 \cdot 5^u$, $CD = 25 \cdot 0^u$ and the inclinations to the horizontal of AB and CD are $67 \cdot 38°$ and $-73 \cdot 74°$ respectively. Then, with $ß = EI$, the pin-end conditions at D are given by:

$$Aß(56 \cdot 5)^3 + Bß(56 \cdot 5)^2 + R_B 37^3/3! + \Delta_B + 9 \times 27^3/3! + R_c 25^3/3! + \Delta_c = 0,$$
$$(2.13.3)$$

5

and
$$6A\text{ß}(56\cdot5)+2B\text{ß}+37R_B+9\times27+25R_c = 0. \qquad (2.13.4)$$

Now for AB, $(a_A)_+ = 0$, since A is a fixed point. An application of $(2.13.1)$ at B then yields:

$$A(19\cdot5)^3+B(19\cdot5)^2+\Delta_B = \cos\alpha\{A(19\cdot5)^3+B(19\cdot5)^2\}, \qquad (2.13.5)$$

and, applying $(2.13.2)$ at B,

$$(a_B)_+ = \sin\alpha\,(A(19\cdot5)^3+B(19\cdot5)^2).$$

At C, $(2.13.1)$ gives

$$(1-\cos\beta)(A\text{ß}(31\cdot5)^3+B\text{ß}(31\cdot5)^2+R_B\times12^3/3!+\text{ß}\Delta_B+9\times2^3/3!)+\Delta_C\text{ß}$$
$$= -\sin\alpha\sin\beta\{A(19\cdot5)^3+B(19\cdot5)^2\}\text{ß}, \qquad (2.13.6)$$

and $(2.13.2)$ gives

$$(a_c)_+ = \sin\alpha\cos\beta(A(19\cdot5)^3+B(19\cdot5)^2)+\sin\beta(A(31\cdot5)^3+B(31\cdot5)^2$$
$$+R_B(12)^3/3!\,\text{ß}+\Delta_B+9\times2^3/3!\,\text{ß})$$
$$= 0. \qquad (2.13.7)$$

($(a_c)_+ = (a_D)_- = 0$ since D is a fixed point.)

Finally, the equilibrium equation necessary because the frame has one kinematic freedom is best obtained by cutting BC free from the frame at B_- and C_+ and taking moments about O as indicated in fig. 2.13.0. Then

$$6A\text{ß}\times18\cdot4+(6A\text{ß}+R_B+R_C+9)\times17\cdot67+(6A\text{ß}\times19\cdot5+2B\text{ß})$$
$$-(6A\text{ß}\times31\cdot5+2B\text{ß}+12R_B+18)+2\times17\cdot0 = 9\times2\cdot94. \qquad (2.13.8)$$

The terms in $(2.13.8)$ arise in turn from moments due to shear forces at B_- and C_+, moments at B and C and finally the 2^u and 9^u loads, respectively.

The equations $(2.13.3–8)$ are the six equations for the six unknowns A, B, R_B, Δ_B, R_C, Δ_C. With a little care the equations are quite easily solved by hand computation. Alternatively, and preferably, by the computer the following results are obtained:

$$\left.\begin{aligned}
A\text{ß} &= 0\cdot004467, \\
B\text{ß} &= 0\cdot230218, \\
R_B &= -1\cdot174609, \\
\Delta_B\text{ß} &= -74\cdot254318, \\
R_C &= -8\cdot060569, \\
\Delta_C\text{ß} &= -83\cdot537033.
\end{aligned}\right\} \qquad (2.13.9)$$

2.13 *Frames with kinematic freedoms II*

It is now a simple matter to construct the S.F. and B.M. diagrams and plot the displaced form (fig. 2.13.2).

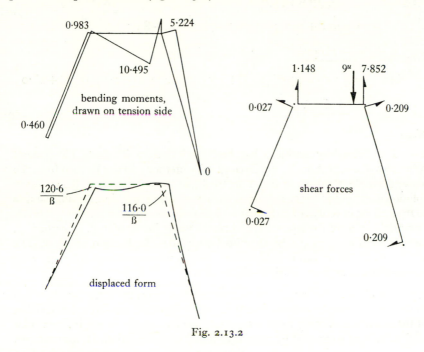

Fig. 2.13.2

For example, $(S.F.)_A = 6A\beta = \underline{0 \cdot 0268^u}$,

$(B.M.)_A = 2B\beta = \underline{0 \cdot 460^u}$,

$(S.F.)_{B+} = 6A\beta + R_B = 0 \cdot 0268 - 1 \cdot 1746$

$\qquad = \underline{-1 \cdot 148^u}$,

which is also the *vertical* reaction at A.

$(B.M.)_B = 6A\beta \times 19 \cdot 5 + 2B\beta,$

$\qquad = \underline{0 \cdot 983^u}.$

$(S.F.)_{c-} = -1 \cdot 148 + W = \underline{7 \cdot 852^u}.$

$(B.M.)_c = \underline{5 \cdot 224^u}.$

$\beta(w)_{B-} = 0 \cdot 004467(19 \cdot 5)^3 + 0 \cdot 2302(19 \cdot 5)^2$

$\qquad = 33 \cdot 1 + 87 \cdot 5 = \underline{120 \cdot 6^u}.$

$\beta(w')_B = 3 \times 0 \cdot 004467(19 \cdot 5)^2 + 2 \times 0 \cdot 2302(19 \cdot 5)$

$\qquad = 5 \cdot 1 + 9 \cdot 0 = \underline{14 \cdot 1^u}.$

$$\text{ß}(w)_{c+} = 0 \cdot 004467(31 \cdot 5)^3 + 0 \cdot 2302(31 \cdot 5)^2 - 1 \cdot 1746(12)^3 - 74 \cdot 2543$$
$$+ 12 - 83 \cdot 5370$$
$$= 139 \cdot 8 + 228 \cdot 5 - 338 \cdot 5 - 74 \cdot 3 + 12 - 83 \cdot 5$$
$$= \underline{-116 \cdot 0^u}.$$

$$\text{ß}(w')_c = 3 \times 0 \cdot 004467(31 \cdot 5)^2 + 2 \times 0 \cdot 2302(31 \cdot 5) - 1 \cdot 1746(72) + 18 \cdot 0$$
$$= \underline{-38 \cdot 7^u}.$$

Notes

2.6   The preceding problem has been included to illustrate the general principles of application for the discontinuity method. As can be seen from the above example, the bulk of the effort to obtain the solution is in the equation solution itself and this is clearly a task for a computing machine. Again, changes to the loading or fixity conditions or variations to the sections chosen, generally affect only some of the equations and can be easily incorporated. In particular, various loading cases can be examined with little duplication of effort. The various load vectors can be formulated and the whole problem solved as a set of equations with multiple right-hand sides.

2.7   The choice of origin clearly has some effect upon the difference between the largest and smallest coefficient in the equations and although the *condition* of the equations is likely to be satisfactory for any choice of origin, particular choices may be suggested by what is required of the solution. If, for example, quantities are required at only one point in the framework for design purposes, this suggests that the origin should be placed at this point. Again, if the origin is placed in the *middle* of the structure, by which is meant *equidistant from the ends*, this will have the effect of minimizing the value of the largest coordinate distance used and hence, in turn, reduce the value of the largest coefficients since these tend to be Ax^3 terms in *displacement* equations. On the whole, though, the origin should be chosen for convenience.

2.14   Simple frameworks III: elastic–plastic

As was seen to be the case with the beam, the collapse load computed from the elastic–plastic calculation is identical with that obtained from the simplified theory, the rigid–plastic theory. The purpose of the elastic–plastic calculations is to supplement this collapse load estimate with estimates for the load at various stages of the load cycle up to collapse and to provide in addition the displacements (the stiffness information) at any stage of the load cycle up to and at the point of collapse.

In some practical situations, displacement limitations required of the

2.14 *Simple frameworks: elastic–plastic*

structure may be reached at loads less than the hinge collapse load. The rigid–plastic theory is then inappropriate and an elastic–plastic calculation must be made.

As an example of this type of calculation we shall examine the displacements produced at collapse in the frame fig. 2.5.0, which in 2.5 was considered from the rigid–plastic standpoint. The expression there obtained for the collapse load and internal force distribution at collapse can now be used as a starting point for the elastic–plastic calculation.

The difference between the elastic and rigid response is clearly that if the material is elastic–plastic, hinges form one at a time as the load is increased and restricted rotation takes place at each. As the last increment of loading to produce collapse is added, the final hinge forms and a mechanism results. In the rigid–plastic case all the hinges can be thought of as forming simultaneously.

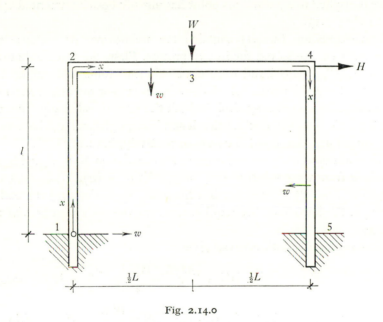

Fig. 2.14.0

We shall adopt the same notation as in fig. 2.5.0, which for convenience is reproduced as fig. 2.14.0. Then if

$$\tfrac{1}{4} < Hl/WL < 1 \tag{2.14.0}$$

the actual mechanism to form will have hinges at 1, 3, 4 and 5.

69

Hence

$$w = \frac{Ax^3}{3!\beta} + \frac{Bx^2}{2!\beta} + \frac{F_2 z_2^3}{3!\beta} + \Delta_2 + \frac{W z_3^3}{3!\beta} + \theta_3 z_3 + \frac{F_4 z_4^3}{3!\beta} + \theta_4 z_4 + \Delta_4. \quad (2.14.1)$$

This displacement description is precisely that for the elastic calculation except for the addition of the $\theta_{3,4}$ terms in anticipation of hinges forming at 3 and 4 during the loading cycle. We shall assume the final hinge to form to be that at 1, although this must be checked (see note 2.9). If the final hinge to form is not at 1 then some rotation will occur at 1 prior to collapse and a Cx term must be included in (2.14.1).

The situation we wish to analyse for is the displaced form, and hence the force distribution and collapse load, as M_1 reaches the value M_p but before any rotation occurs at hinge site 1.

Note the absence of any term in H in (2.14.1). The reason is because the force is applied to a joint. This point has already been commented upon in 2.12 and 2.13.

The expression (2.14.1) contains ten unknowns. At collapse the unknowns can be solved for in two groups. First, the force unknowns A, B, F_2, W, F_4 are found from equilibrium considerations alone as a repeat of the rigid–plastic calculation. In order that the unknowns may be denoted by common symbols in both the present calculation and the rigid–plastic calculation of 2.5 the denominators $3!\beta$ and $2!\beta$ have been introduced into the A and B terms, respectively, in (2.14.1).

Then as the second and final stage in the calculation Δ_2, θ_3, θ_4 and Δ_4 are obtained from the conditions $(w)_{2+} = 0$, $-(w)_{2-} = (w)_{4+}$, $(w)_{4-} = 0$ and $(w)_5 = 0$. The latter condition is a boundary condition and the earlier *three* are the inextensibility requirements for the *three* members of the frame.

Now the rigid–plastic analysis gives

$$B = M_p, \quad F_2 L = 4M_p - W_c L + \frac{4M_p L}{l} - \frac{W_c L^2}{2l}, \quad \frac{W_c L}{2} + H_c l = 6M_p,$$

$$F_4 = -(H_c + 2A + W_c + F_2), \quad Al = \frac{W_c L}{2} - 4M_p. \quad (2.14.2)$$

Once a shape of frame (l/L) and the ratio of W_c/H_c are chosen, the collapse loads (W_c, H_c) and A, F_2, F_4 follow from (2.14.2). Then from the remaining displacement conditions

$$(w)_{2+} = \frac{Al^3}{6\beta} + \frac{Bl^2}{2\beta} + \Delta_2 = 0, \quad (2.14.3)$$

$-(w)_{2-} = (w)_{4+}$ gives

$$-\left(\frac{Al^3}{6\beta} + \frac{Bl^2}{2\beta}\right) = \frac{A(l+L)^3}{6\beta} + \frac{B(l+L)^2}{2\beta} + \frac{F_2 L^3}{6\beta} + \Delta_2 + \frac{W_c L^3}{48\beta} + \theta_3 \frac{L}{2} + \Delta_4,$$

(2.14.4)

$$(w)_{4-} = \frac{A(l+L)^3}{6\beta} + \frac{B(l+L)^3}{2\beta} + \frac{F_2 L^3}{6\beta} + \Delta_2 + \frac{WL^3}{48\beta} + \theta_3 \frac{L}{2} = 0, \quad (2.14.5)$$

$$(w)_5 = \frac{A(2l+L)^3}{6\beta} + \frac{B(2l+L)^2}{2\beta} + \frac{F_2(l+L)^3}{6\beta} + \Delta_2 + \frac{W(\frac{1}{2}L+l)^3}{6\beta} + \theta_3(\frac{1}{2}L+l)$$

$$+ \frac{F_4 l^3}{6\beta} + \theta_4 l + \Delta_4 = 0. \quad (2.14.6)$$

From the first three equations (2.14.3, 4, 5) it is clear that

$$\Delta_2 = \Delta_4. \tag{2.14.7}$$

With particular ratios of l/L, W_c/H_c, then (2.14.2–7) are easily solved. For example, $l = L$, $W_c = 2H_c$, from (2.14.2–7)

$$Wl = 6M_p, \quad Al = -M_p, \quad w_3\beta = \tfrac{5}{24}M_p l^2, \tag{2.14.8}$$

and the sidesway displacement Δ is

$$\Delta\beta = M_p l^2/3. \tag{2.14.9}$$

Notes

2.8   The history of moments and displacements at stages of the loading after first yield but before collapse can be studied directly from (2.14.1) by deletion of the appropriate θ terms and full plastic moment equations. But this will then mean that all the unknowns must be solved for simultaneously and not as a force group followed by a dependent displacement group.

2.9   The choice of point 1 as the site of the last hinge to form needs checking. It is strongly reinforced from a knowledge of the elastic distribution of bending moments which is $(17, -4, -28, 32, -31) \times (WL/84)$ at the points 1–5 respectively ($L = l$, $W = 2H$). As a first guess, the order of hinge formation is likely to be therefore 4, 5, 3 and 1, respectively. This can be shown to be correct by calculation. In more complicated examples, however, other checks should be applied to confirm such a choice based on a knowledge of the elastic distribution. The limit theorems are the most useful tool in such checks.

2.15 More extensive frameworks: two storey elastic: I symmetric loading

The treatment of rigid joints by means of discontinuity solutions ensures that any continuity conditions at a joint are automatically taken care of. This is the primary usefulness of the present approach. Another situation in which the continuity across a junction can be handled compactly is where a column is continuous through from floor to floor and a single x coordinate is used to describe all points on this column.

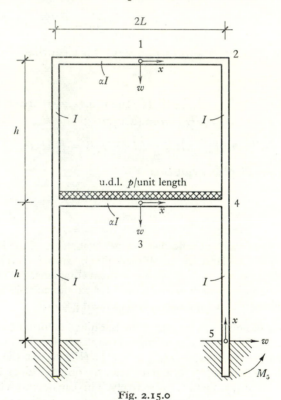

Fig. 2.15.0

Consider the two storey, single bay rigid jointed elastic frame shown in fig. 2.15.0. The minimum number of coordinate systems necessary in this case is two. More may be used if desired and in this case three have been chosen, as indicated in fig. 2.15.0, in order to exploit to the full the symmetry of the structure.

Then, for 1–2

$$w = b(x^2 - L^2), \tag{2.15.0}$$

for 3–4
$$w = \beta(x^2 - L^2) + \frac{p(x^4 - L^4)}{4!\,\alpha\beta},$$

(2.15.1)

for 2–5
$$w = Ax^2(x - h) + \frac{M.z^2}{2!\,\beta} + \frac{T.z^3}{3!\,\beta},$$

(2.15.2)

where $z = x - h$.

Now (2.15.0–2) clearly already incorporate the obvious symmetry conditions at 1 and 3 together with the fixity conditions at 5 and the zero displacement conditions at 2 and at 4 for 1–2 and 3–4.

There remain five unknowns b, β, A, M and T which are obtained from the following five conditions.

Slope continuity at 4
$$A.h^2 = 2L\beta + \frac{pL^3}{3!\,\alpha\beta}.$$

(2.15.3)

Moment continuity at 4
$$M = 2\alpha\beta\beta + \frac{pL^2}{2}.$$

(2.15.4)

Slope continuity at 2
$$2Lb = 8h^2A + \frac{Mh}{\beta} + \frac{Th^2}{2\beta}.$$

(2.15.5)

Moment continuity at 2
$$-2\alpha\beta b = 10h\beta A + M + Th,$$

(2.15.6)

and finally, displacement zero at 2 for 5–2
$$4h^3A + \frac{Mh^2}{2\beta} + \frac{Th^3}{2!\,\beta} = 0.$$

(2.15.7)

The overall equilibrium balances for the beams are automatically satisfied by symmetry in the present case.

Eliminating T from (2.15.5–7)
$$\left. \begin{array}{l} -4(L/h)\,\beta b = 8h\beta A + M, \\ 2\alpha\beta b = 14h\beta A + 2M. \end{array} \right\}$$

(2.15.8)

Now eliminate b and use (2.15.3–4) when
$$\beta = -\frac{pL^2}{12\alpha\beta}\,\frac{28L^2 + 20\alpha Lh + 3\alpha^2h^2}{28L^2 + 12\alpha Lh + \alpha^2h^2}.$$

(2.15.9)

Also
$$A = \frac{2L\beta}{h^2} + \frac{pL^3}{6\alpha\beta h^2}$$

$$= -\frac{pL^3}{3\beta h}\,\frac{4L + \alpha h}{28L^2 + 12\alpha Lh + \alpha^2h^2}.$$

(2.15.10)

We are now able to evaluate any quantity of interest. For example, M_5 as indicated on fig. 2.15.0 is obtained as

$$M_5 = -2\text{ß}hA = \frac{2pL^3}{3}\frac{4L+\alpha h}{28L^2+12\alpha Lh+\alpha^2h^2}, \qquad (2.15.11)$$

and the displacement at 3 as

$$w_3 = -\beta L^2 - \frac{pL^4}{4!\,\alpha\text{ß}} = \frac{pL^4}{4!\,\alpha\text{ß}}\frac{28L^2+28\alpha Lh+5\alpha^2h^2}{28L^2+12\alpha Lh+\alpha^2h^2}. \qquad (2.15.12)$$

As has been shown earlier, addition of appropriate θz terms will allow change of the above elastic analysis into an elastic–plastic analysis; alternatively if the shape functions are twice differentiated with respect to x then multiplied by $\text{ß} = EI$, the section stiffness, the resulting expressions are the bending moment distributions on which a rigid–plastic analysis can be made.

It is also of interest to note that all the derived quantities concerned with the framework are rational functions of the parameter $\alpha h/L$. An elementary dimensional analysis would indicate that α and h/L are dimensionless groups likely to be of interest but only the analysis above, or equivalent, can reveal the additional and very useful information that the *product* $\alpha h/L$, and not the individual ratios, is of primary importance.

This information is important for example in model analysis where now it can be seen that, provided α can be made to vary over a wide enough range in the model, all two storey frames with various α's and various h/L's can be studied from a single model with say $h/L = 1$ and α large to begin with but made smaller by paring down the beam size as the testing proceeds. Then information relating to $h/L = \frac{1}{2}$ and $\alpha = 2$, for example, can be obtained from $\alpha = 1$ and $h/L = 1$, namely the model proposed.

Example

With the same frame as in fig. 2.15.0, but with the upper rather than the lower storey loaded, derive the following results. The notation is as in the previous example with, in addition,

$$f = f(\alpha h/L) = 28L^2 + 12\alpha Lh + \alpha^2h^2.$$

Then
$$A = 2pL^4/3\text{ß}hf,$$
$$(M_5)_4 = -4pL^4/3f,$$
$$M = 8pL^4/3f,$$
$$(M_4)_5 = 8pL^4/3f$$
$$= 2|(M_5)_4|.$$

Here $(M_\alpha)_\beta$ is used to denote the moment in member $\alpha\beta$ at end α.

2.16 More extensive frameworks: two storey elastic II: antisymmetric loading

Consider again the same frame as in fig. 2.15.0 but now subject to beam level horizontal forces, H/storey, producing an antisymmetric system. The loading could be from wind forces or seismic action. Other, non-uniform, antisymmetric loads can easily be incorporated. This anti-symmetry ensures that the mid points of the beams are points of zero displacement and bending moment, i.e. $w = w'' = 0$ (where symmetry

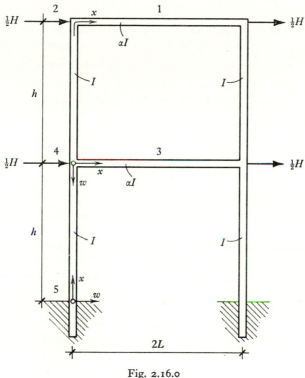

Fig. 2.16.0

ensures $w' = w''' = 0$). In this example only two coordinate systems will be used as indicated in fig. 2.16.0. They will be an origin at 5 for 5–4–2–1 and an origin at 4 for 4–3.

Then, for 5–4–2–1

$$w = Ax^3 + Bx^2 + \frac{Hz_4^3}{2 \times 3!\,\beta} + \frac{M_4 z_4^2}{2!\,\beta} + \frac{F_2 z_2^3}{3!\,\alpha\beta} + \frac{M_2 z_2^2}{2!\,\alpha\beta} + \Delta_2, \quad (2.16.0)$$

where $z_4 = x - h$, $z_2 = x - 2h$, and for 4–3

$$w = ax^3 + bx^2 + cx. \qquad (2.16.1)$$

There are nine unknowns. Physically M_4, F_2, M_2 are discontinuities in moment and shear experienced by 5–4–2–1; Δ_2 is the discontinuity in displacement, in fact the sidesway displacement at 2. A, B; a, b, c have their usual interpretations as to actions at an origin.

The frame has two degrees of kinematic freedom but only one overall equilibrium equation need be written down since continuity through 4 for 5–4–2 has allowed us to include the term in H at 4 which usually necessitates writing an additional equation.

The known conditions at either origin have already been introduced. The remaining nine conditions are: zero displacement and moment at 1 and 3 (four equations); zero displacement at 2 for 2^+ and moment continuity (two equations); slope and moment continuity at 4 (two equations) and finally a shear balance for the upper storey (one equation, making nine in all).

In full these equations are:

$$A(2h+L)^3 + B(2h+L)^2 + \frac{H(h+L)^3}{12\beta} + \frac{M_4(h+L)^2}{2\beta}$$

$$+ \frac{F_2 L^3}{6\alpha\beta} + \frac{M_2 L^2}{2\alpha\beta} + \Delta_2 = 0, \quad (2.16.2)$$

$$6A(2h+L)\beta + 2B\beta + \frac{H(h+L)}{2} + M_4 + \frac{F_2 L}{\alpha} + \frac{M_2}{\alpha} = 0, \quad (2.16.3)$$

$$aL^3 + bL^2 + cL = 0, \qquad (2.16.4)$$

$$3aL + b = 0, \qquad (2.16.5)$$

$$A(2h)^3 + B(2h)^2 + \frac{Hh^2}{12\beta} + \frac{M_4 h^2}{2\beta} + \Delta_2 = 0, \qquad (2.16.6)$$

$$M_2 = (1-\alpha)\beta\left(12AL + 2B + \frac{Hh}{2\beta} + \frac{M_4}{\beta}\right), \qquad (2.16.7)$$

$$c = 3Ah^2 + 2Bh, \qquad (2.16.8)$$

$$M_4 = -2b\alpha\beta, \qquad (2.16.9)$$

and, finally, the shear balance,

$$\left(6A + \frac{H}{2\beta}\right)\beta = -\frac{H}{2} \quad \text{or} \quad 6A\beta = -H. \qquad (2.16.10)$$

These nine equations are quite easily solved to give, in particular B, thence the actions at the column foot and hence all the actions in easy fashion.

Thus:
$$(M_5)_4 = 2BEI = \frac{3Hh}{2} \frac{L^2 + 4 \cdot 5 \alpha Lh + 3\alpha^2 h^2}{L^2 + 9\alpha Lh + 9\alpha^2 h^2}. \qquad (2.16.11)$$

This example is a simplified version of a frame of considerable practical importance, the single bay multi-storey frame. We shall delay any further discussion of the present class of frame until we consider this particular multi-storey frame in detail, later in this chapter.

Once again, the results of the calculation show the dependent quantities to be rational functions of the parameter $\alpha h/L$. The complete expression for $(M_5)_4$, for example, consists of $f(\alpha h/L)$ multiplying Hh. Now the structure is basically a cantilever with axis vertical and 'length' $2h$. Hence we would expect that the moment at the root, that is point 5, is of the form (coeff. $(\alpha h/L)$) . Hh, which it is. Referring back to 2.15 where the structure is basically a beam spanning $2L$, the bending moments we would on this reasoning expect to be of the form coeff. $(\alpha h/L)$. L^2, where in either case the expression 'coeff. $(\alpha h/L)$' is a rational function of $\alpha h/L$ only. It can be seen there too that the results are of this form.

2.17 Conclusion to discussion of simple frameworks

As with beams, the main theme of the present chapter has thus far been illustrated with comparatively simple examples. The object has been to illustrate the close connexion between beam bending problems and frameworks whether they be treated from the rigid–plastic, elastic or elastic–plastic view point. More difficult examples, and especially examples which demand use of a computer to solve the resulting equations, can be found in the relevant items noted in the references.

Our subject has been treated entirely from the standpoint of discontinuity methods. This is a novel feature but one we think justified. In the literature other methods are used to deal with similar problems. These methods on the one hand are based on energy principles and lead to the so-called force method using forces (redundants) as unknowns, or deal with the single member differential equation with no particular integrals and lead to the so-called displacement method using joint displacements and rotations as unknowns. The present approach runs an intermediate course and uses a selection of both forces and displacements as unknowns. It shares in large measure the advantage which the displacement method has over the force method in presenting a readily definable algorithm for formulating the problem equations while at the same time gives a method suitable for hand computation. The displacement method in contrast is uncompromisingly machine oriented and unsuitable for hand computation except in the simplest cases.

In a later treatment it is planned to restate the discontinuity methods in a form more immediately suitable for machine computation. This will require use of matrices. In this introductory treatment matrices have been excluded as a calculation tool since it is felt some understanding of structural problems should be gained from hand computations before the matrix treatment and machine computation are introduced.

2.18 Further exercises

The frames 1–7 below are assumed to be operating in the elastic regime.

(1) A uniform elastic member of length 15 units and fixed in position and direction at each end has a moment value 5 force × length units applied at the third point. Plot the B.M. diagram.

(2) Two uniform elastic members each of length 5 units are rigidly jointed so as to span 8 units, with the apex 3 units above the level abutments to which the framework is pinned. A vertical load of 5 units is applied to the mid-point of one member. Plot the B.M. diagram.

The following four examples concern a rigidly jointed elastic, symmetrical rectangular portal of span $2a$, height h and with (EI) Beam $= \alpha \times (EI)$ column $= \alpha \beta$.

(3) For the above frame, if pinned to the foundations and subject to a horizontal beam level force, H, show that the displacement in the direction of the force H, the sway displacement, is given by

$$\Delta = \frac{Hh^2(h+(a/\alpha))}{6\beta}.$$

(4) As in (3) above, but fixed to the foundations, show that the sway displacement Δ is given by

$$\Delta = \frac{Hh^3(3h+(4a/\alpha))}{24\beta(3h+(a/\alpha))}$$

and

$$M = \frac{Hh}{4}\frac{3h+(2a/\alpha)}{3h+(a/\alpha)},$$

where M is the foundation column moment.

(5) The frame as in (3), that is pinned to the foundation, but loaded with a symmetrical beam load such as to produce end moments of F in a beam of

2.18 *Further exercises*

length $2a$ when held fixed against rotation at the ends. In the method of moment distribution $\pm F$ are called the fixed end moments. Then if

$$p+q = 1 \quad \text{with} \quad p \propto \frac{3}{4}\frac{\beta}{h}, \quad q \propto \frac{\alpha\beta}{2a}$$

show that the moment (M) at the beam/column junction is given by

$$M = \left(\frac{2p}{2-q}\right)F.$$

The p and q so defined are called the 'distribution factors'.

(6) For the frame loaded as in (5) above but fixed to the foundations, show that if $p \propto \beta/h$, $q \propto \alpha\beta/2a$, $p+q = 1$,

then
$$M = \left(\frac{2p}{2-q}\right)F \quad \text{as in 5},$$

and the foundation column moment $= |\frac{1}{2}M|$.

(7) Show that if the beam in (5) or (6) is loaded with a u.d.l. p/unit length then
$$F = (pa^2)/3.$$

If loaded with a central vertical beam load W show that

$$F = (Wa^2)/2.$$

(8) The following rigid–plastic problem is discussed by Heyman (2.26). The frame is shown in fig. 2.18.0. Heyman's units of length are inches and of force, kips (that is 1,000 lbf units). The boxed figures shown adjacent to the members are the M_p values in Kip inches $\div$ 10^3 units. It is required to evaluate the load factor against collapse. Namely, what is the smallest multiple by which all the loads must be multiplied in order to produce collapse in the framework according the usual rigid–plastic requirements of no instability etc.

Now a description of the bending moment in the frame in the spirit of the discussion in the present chapter is, for 1–5, 11, 6–10, with origin at 1

$$M = Cx + D + \Delta_2 + \Delta_3 + \Delta_4 + \Delta_7 + \Delta_8 + \Delta_9 + \sum_{2}^{9}\phi_i z_i + 30\lambda z_{11};$$

and for a typical beam, 4, 12, 7 say, with origin at 4,

$$M = C_4 x + d_4 + 30\lambda z_{12}.$$

The C, D; C_i, d_i are complementary function constants and the Δ_i, ϕ_i are typical moment and S.F. discontinuities.

Investigate possible mechanisms and show that the most critical one, that is the one which for chosen hinge positions produces a minimum value for the load factor λ, has hinges at 1, 11, 6_+, 10 and beam hinges at 12, 7; 13, 8; 14, 9 for which, with

$$\pi_1 = \sum_{2}^{5}\phi_i,$$

$$\pi_2 = \sum_{6}^{9}\phi_i,$$

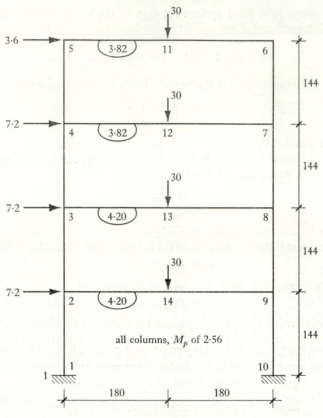

Fig. 2.18.0

then
$$d_4 = -\Delta_4, \quad \text{etc.,}$$

$$\pi_1 - \pi_2 = 186.7 - 56.6\lambda,$$

$$\pi_1 + \pi_2 = -(2C + 55.2\lambda),$$

$$\pi_1 = 35.5 - 30\lambda - C.$$

Hence
$$\lambda = 2.23.$$

If the mechanism has hinges in the columns at 1, 2_+, 9_+, 5 and in the beams at 14 and 9 show that
$$\lambda = 2.53.$$

Or again, for hinges in the columns only at 1, 2_-; 9_+, 10 show that
$$\lambda = 2.82.$$

A check of the first mechanism will show it to be the correct one.

2.19 Solutions to exercises

(1) Here

$$w = Ax^3 + Bx^2 + \frac{Mz^2}{2!\ss} \quad (z = x - 5).$$

From the conditions at 3

$$A \times 15^3 + B \times 15^2 + \frac{M \times 10^2}{2\ss} = 0,$$

$$3 \times A \times 15^2 + 2B \times 15 + \frac{M \times 10}{\ss} = 0.$$

Hence $\quad B = 0, \quad A = \dfrac{2M}{135EI},$

$$M_{2-} = 6 \times A \times EI \times 5 = \tfrac{4}{9}M \quad (= 2 \cdot 2 \text{ for } M = 5).$$

(2) No kinematic freedoms, $\Delta_{2+} = \Delta_{2-} = 0.$ Hence

$$w = Ax^3 + Cx + \frac{5\cos\alpha \times z_4^3}{3!\ss} + \frac{Fz_2^3}{3!\ss}.$$

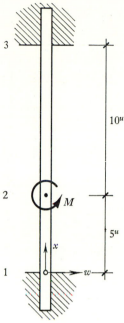

Fig. 2.19.0

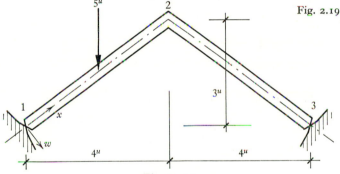

Fig. 2.19.1

Unknowns A, C, F. Conditions required are $w_2 = 0 = w_3$, $w_3'' = 0$. Hence

$$A \times 5^3 + C \times 5 + \frac{4}{6}\frac{(2 \cdot 5)^3}{\ss} = 0,$$

$$A \times 10^3 + C \times 10 + \frac{4}{6}\frac{(7 \cdot 5)^3}{\ss} + \frac{F \times 5^3}{6\ss} = 0,$$

$$A \times 60 + \frac{4 \times 7 \cdot 5}{\ss} + \frac{F \times 5}{\ss} = 0.$$

Solving, $6AEI = $ shear at $1 = -1 \cdot 625$, and $R_1 = 3 \cdot 75$, $R_2 = 1 \cdot 25$, $H = 2 \cdot 29$.

(3) Here
$$w = Ax^3 + Cx + \frac{F(z_2)^3}{2!\alpha\beta} + \frac{M(z_2)^2}{3!\alpha\beta} + \Delta_2.$$

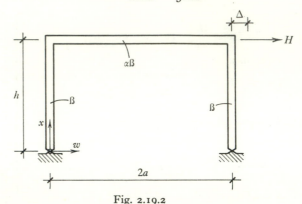

Fig. 2.19.2

The conditions are:

(a) $6A\beta = -H/2$, shear at 1.

(b) $6Ah\alpha\beta + M = 6Ah\beta$, moment continuity at 2.

(c) $Ah^3 + Ch + \Delta_2 = 0$, inextensibility of column.

(d) $A(a+h)^3 + C(a+h) + \dfrac{Fa^3}{6\alpha\beta} + \dfrac{Ma^2}{2\alpha\beta} + \Delta_2 = 0,$ ⎫ Antisymmetry condition

(e) $6A(a+h)\alpha\beta + Fa + M = 0.$ ⎬ condition $w_3 = w_3'' = 0.$

These are the five equations in the five unknowns, A, C, F, M, Δ_2. Then, from (d), (e):
$$6A(a+h)\alpha\beta(2ah + h^2) + 6C(a+h)\alpha\beta + 2Ma^2 + 6\Delta_2\alpha\beta = 0.$$

Using (c),
$$6A(a+h)\alpha\beta \cdot 2ah + 2Ma^2 + 6\Delta_2\alpha\beta\left(1 - \frac{a+h}{h}\right) = 0.$$

Whence, using (a) and (b):
$$6\Delta_2\beta = -h^2(a+h)H + \frac{ah}{\alpha}Hh(\alpha-1),$$

$$= -Hh^2\left(a + h + \left(\frac{1-\alpha}{\alpha}\right)a\right) = -Hh^2\left(h + \frac{a}{\alpha}\right).$$

Thence
$$\Delta = -\Delta_2 = \frac{Hh^2(h + a/\alpha)}{6\beta}.$$

Note

The displaced form includes a 'moment' discontinuity term in M because of the change in section stiffness through the joint 2. The companion 'shear' term has been absorbed into F which includes also a contribution from the change in direction of the members at 2.

2.19 Solutions to exercises

The remaining exercises (4)–(7) are very similar in character to exercise (3). Details are omitted.

(8) This exercise is more quickly solved, if only the collapse load is required, by the work equation derived in (1.4). But by the present methods the results follow from consideration of the frame at collapse as shown in fig. 2.19.3.

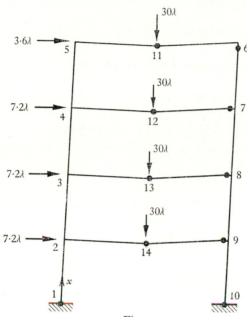

Fig. 2.19.3

Mode I

The unknowns are λ, C, D, Δ_{2-4}, Δ_{7-9}, ϕ_{2-9} together with an additional unknown for each beam not already included; 20 unknowns in all.

The equations necessary to solve are obtained in a systematic (if lengthy) fashion from: conditions on moments being full plastic at hinges (13 equations), and shear balances for each storey (4 equations) making 17 equations in all.

Hence it is clear that not all the unknowns can be solved for explicitly. At collapse in Mode I the frame is three times redundant. Those which can be easily found are the Δ_{7-9} and D.

(1–4) Clearly

$$\Delta_7 = 3 \cdot 82 \times 10^3, \quad \Delta_8 = 4 \cdot 20 \times 10^3, \quad \Delta_{10} = 4 \cdot 20 \times 10^3, \quad D = 2 \cdot 56 \times 10^3$$

although these results are mechanism dependent. For a typical beam 4–12–7:

$$M = C_4 x + d_4 + 30 \times z_{12};$$

whence

$$M_4 = d_4 = -\Delta_4.$$

Similarly,

$$d_3 = -\Delta_3, \quad d_2 = -\Delta_2.$$

83

Forces, typically C, C_{2-4} will be measured in Kip units, and moments, typically $\Delta_{2,4}$ will be measured in Kip inch units.

Then a typical shear balance equation, obtained in this case by cutting the frame free at ground level is

$$(5) \quad C + \left(C + \sum_2^9 \phi_i + 30\lambda\right) = -(3 \times 7\cdot2 + 3\cdot6)\,\lambda.$$

For the frame cut free through the second storey

$$(6) \quad C + \phi_2 + \left(C + \sum_2^8 \phi_i + 30\lambda\right) = -(2 \times 7\cdot2 + 3\cdot6)\,\lambda.$$

The remaining two such equations are

$$(7) \quad C + \phi_2 + \phi_3 + \left(C + \sum_2^7 \phi_i + 30\lambda\right) = -(7\cdot2 + 3\cdot6)\,\lambda.$$

$$(8) \quad C + \sum_2^4 \phi_i + \left(C + \sum_2^6 \phi_i + 30\lambda\right) = -3\cdot6\lambda.$$

These four equations are not mechanism dependent and clearly possess a number of symmetries.

By simple summations we see that

$$\phi_2 - \phi_9 = \phi_3 - \phi_8 = \phi_4 - \phi_7 = 7\cdot2\lambda, \tag{a}$$

Also, from (5),
$$2C + \pi_1 + \pi_2 = -55\cdot2\lambda, \tag{b}$$

one of the results sought.

Continuing, for beam 4–12–7:

$$(9) \quad M_{12} = -3\cdot82 \times 10^3 = C_4 \times 180 - \Delta_4,$$

$$(10) \quad M_7 = 3\cdot82 \times 10^3 = C_4 \times 360 - \Delta_4 + 30 \times \lambda \times 180,$$

whence
$$\Delta_4 = 11\cdot46 \times 10^3 - 5400 \times \lambda,$$

and
$$C_4 = 42\cdot4 - 30 \times \lambda.$$

In a similar manner,

$$(11)–(12) \quad \Delta_3 = 12\cdot60 \times 10^3 - 5400 \times \lambda = \Delta_2,$$

$$(13)–(14) \quad C_3 = 46\cdot7 - 30 \times \lambda \qquad = C_2.$$

The remaining three equations express that the moments at 11, 6 and 10 are fully plastic:

$$(15) \quad (144 \times 4 + 180)\,C + 2\cdot56 \times 10^3 + \sum_2^4 \Delta_i + \phi_2(3 \times 144 + 180)$$

$$+ \phi_3(2 \times 144 + 180) + \phi_4(144 + 180) + \phi_5 \times 180 = -3\cdot82 \times 10^3.$$

(16) $(144 \times 4 + 360)\,C + 2 \cdot 56 \times 10^3 + \sum_2 \Delta_i + \phi_2(3 \times 144 + 360)$

$$+ \phi_3(2 \times 144 + 360) + \phi_4(144 + 360) + \phi_5(360) + 30 \times 180 \times \lambda$$

$$= 2 \cdot 56 \times 10^3.$$

(17) $(126 \times 12)\,C + 2 \cdot 56 \times 10^3 + \sum_2^4 \Delta_i + \sum_7^9 \Delta_i + \phi_2(114 \times 12) + \phi_3(102 \times 12)$

$$+ \phi_4(90 \times 12) + \phi_5(78 \times 12) + \phi_6(48 \times 12) + \phi_7(36 \times 12)$$

$$+ \phi_8(24 \times 12) + \phi_9(12 \times 12) + 30\lambda(180 + 4 \times 144) = -2 \cdot 56 \times 10^3.$$

From a difference of (15) and (16)

$$180C + \pi_1 \times 180 + 30 \times \lambda \times 180 = +6 \cdot 38 \times 10^3$$

or $$C + \pi_1 + 30\lambda = +35 \cdot 5, \qquad (c)$$

another of the results sought.
We have already that

$$\sum_7^9 \Delta_i = (3 \cdot 82 + 8 \cdot 40) \times 10^3$$

$$= 12 \cdot 22 \times 10^3 \quad \text{Kip inches}$$

and $$\sum_2^4 \Delta_i = -16200\lambda + 36 \cdot 66 \times 10^3.$$

Now 5–8 summed give

$$8C + 7\phi_2 + 6\phi_3 + 5\phi_4 + 4\phi_5 + 4\phi_6 + 3\phi_7 + 2\phi_8 + \phi_9 + 120\lambda$$

$$= -(6 \times 7 \cdot 2 + 4 \times 3 \cdot 6)\,\lambda.$$

If this equation is multiplied by 144 and subtracted from (17) then

$$30 \times 12C + \sum_2^4 \Delta_i + \sum_7^9 \Delta_i + 30 \times 12\pi_1 - (30 \times 180 - 57 \cdot 6 \times 144)\,\lambda = -5 \cdot 12 \times 10^3,$$

or $$C + \pi_1 = 150 + 53\lambda, \qquad (d)$$

whence from (c) and (d) $$\lambda_{\mathrm{I}} = 2 \cdot 23.$$

Mode II

In a real problem, alternative modes would need to be investigated, for example the mode shown in fig. 2.19.4. Now

$$M = Cx + 2 \cdot 56 \times 10^3 + 30\lambda z_{14} + \phi_2 z_2 + \phi_9 z_9 + \Delta_2 + \Delta_9,$$

with the coordinate measured only around the lower storey. There are six unknowns.
Now the conditions are as follows. Shear balance:

(1) $2C + 30\lambda + \phi_2 + \phi_9 = -25 \cdot 2\lambda.$

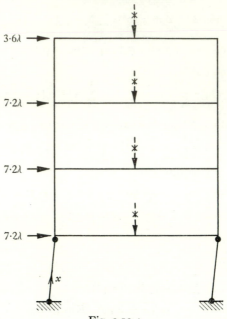

Fig. 2.19.4

Moment at $2-$:

(2) $C \times 144 + 2 \cdot 56 \times 10^3 = -2 \cdot 56 \times 10^3$, $C = \dfrac{-5 \cdot 12 \times 10^3}{144}$.   (a)

Moment at $9+$:

(3) $C(144 + 360) + 2 \cdot 56 \times 10^3 + 30 \times \lambda \times 180 + \phi_2 \times 360 + \Delta_2 + \Delta_9$

$= 2 \cdot 56 \times 10^3$.

Moment at 10:

(4) $C(288 + 360) + 2 \cdot 56 \times 10^3 + 30\lambda(180 + 144) + \phi_2(360 + 144) + \phi_9(144)$

$+ \Delta_2 + \Delta_9 = -2 \cdot 56 \times 10^3$.

Then $\phi_2 + \phi_9 = -30\lambda$ (b), and by difference of 3 and 4

$$C \times 144 + 30 \times \lambda \times 144 + (\phi_2 + \phi_9) \, 144 = -5 \cdot 12 \times 10^3,$$ (c)

whence (a), (b) (c) give $\lambda_{\text{II}} = 2 \cdot 82$.

Mode III

For an upper beam mode fig. 2.19.5:

$$M = Cx - 3 \cdot 82 \times 10^3.$$

But $C = 15\lambda =$ shear at origin, 11.

Hence at 6: $2 \cdot 56 \times 10^3 = 15 \times \lambda \times 180 - 3 \cdot 82 \times 10^3.$

$$\lambda_{\text{III}} = \frac{6 \cdot 38 \times 10^3}{15 \times 180} = 2 \cdot 36.$$

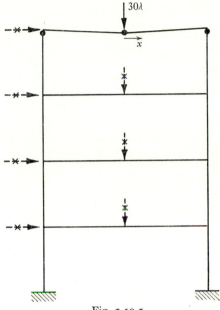

Fig. 2.19.5

Mode IV

The final mode we shall investigate is shown in fig. 2.19.6. Here

$$M = Cx + D + \Delta_2 + \phi_2 z_2 + 30\lambda z_{14} + \Delta_9 + \phi_9 z_9.$$

In this case, all seven unknowns can be solved for from the following.

(1) A shear balance is as for Mode II.
(2) In addition, clearly $D = 2 \cdot 56 \times 10^3$ being the moment at the origin 1.
(3) And $\Delta_2 = 2 \cdot 56 \times 10^3$, being the step change in the moment at 2.
(4) At point 14; $M = M_p$ or $324C + 180\phi_2 = -9 \cdot 32 \times 10^3$.
(5) At $9-$, $504C + 30 \times 180 \times \lambda + 360\phi_2 = -920$, eliminating ϕ_2,

$$144C = -17 \cdot 72 \times 10^3 + 5400\lambda.$$

(6) At $9+$, $504C + 360\phi_2 + \Delta_9 + 30 \times 180 \times \lambda = -2 \cdot 56 \times 10^3$, whence

$$\Delta_9 = 1640, \quad \phi_2 = 170 - 67 \cdot 5\lambda.$$

(7) Finally at 5,

$$(504 + 144)C + (360 + 144)\phi_2 + \Delta_9 + 30\lambda(180 + 144) + \phi_9(144)$$
$$= -(5 \cdot 12 + 2 \cdot 56)10^3.$$

Taking differences

$$C + \phi_2 + 30\lambda + \phi_4 = \frac{-5 \cdot 12 \times 10^3}{144}.$$

Hence $\qquad C = \dfrac{5 \cdot 12 \times 10^3}{144} - 25 \cdot 2\lambda, \quad$ using the shear equation

$$= \frac{-17 \cdot 72}{144} 10^3 + \frac{5400\lambda}{144}$$

from (3) and (4). Hence

$$62 \cdot 7\lambda = 158 \cdot 5, \quad \lambda_{IV} = 2 \cdot 53.$$

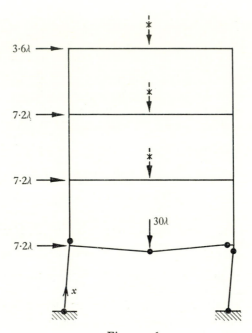

Fig. 2.19.6

It is concluded that Mode I is the most likely one to occur since $\lambda_I < \lambda_{II}$, λ_{III} or λ_{IV}. Before it is finally concluded that λ_I corresponds to the correct mode it must be shown that there exists a possible B.M. distribution with $M = M_p$ at the hinge sites chosen and $|M| < M_p$ elsewhere. Such a check is known as the *static check*.

The frame collapsing in Mode I is three times indeterminate (redundant). We shall choose M_{7-}, M_{8-}, M_{9-} as three (redundant) moments in terms of which, and the load, the B.M. at all points can be written down.

2.19 Solutions to exercises

Now it is straightforward to show that

$$M_{7-} - M_{4+} = -0\cdot43 \times 10^3,$$
$$M_{8-} - M_{3+} = -0\cdot69 \times 10^3,$$
$$M_{9-} - M_{2+} = -1\cdot74 \times 10^3 \quad \text{(Kip inch units)}$$

and

$$\Delta_2 = \Delta_3 = (12\cdot60 - 12\cdot10) \times 10^3 = 0\cdot50 \times 10^3,$$
$$\Delta_4 = (11\cdot46 - 12\cdot10) \times 10^3 = -0\cdot64 \times 10^3 \quad \text{(Kip inch units)}.$$

Thence the beam moments at the column junctions 2, 3, 4 are less than yield. Again

$$-M_{2-} = M_{2+} - \Delta_2 = M_{9-} + 1\cdot24 \times 10^3,$$
$$-M_{3-} = M_{3+} - \Delta_3 = M_{8-} + 0\cdot19 \times 10^3,$$
$$-M_{4-} = M_{4+} - \Delta_4 = M_{7-} + 1\cdot07 \times 10^3.$$

For no yield at $2-$, $3-$, $4-$

$$-3\cdot80 < M_{9-}/10^3 < 1\cdot32,$$
$$-2\cdot75 < M_{8-}/10^3 < 2\cdot37,$$
$$-3\cdot63 < M_{7-}/10^3 < 1\cdot49.$$

For equilibrium at 7, 8, 9 we have

$$M_{7+} + M_{7-} + 3\cdot82 \times 10^3 = 0,$$
$$M_{8+} + M_{8-} + 4\cdot20 \times 10^3 = 0,$$
$$M_{9+} + M_{9-} + 4\cdot20 \times 10^3 = 0.$$

Hence, for no yield at $7+$, $8+$, $9+$

$$-6\cdot38 < M_{7-}/10^3 < -1\cdot26,$$
$$-6\cdot76 < M_{8-}/10^3 < -1\cdot64,$$
$$-6\cdot76 < M_{9-}/10^3 < -1\cdot64.$$

There remains joint 5. Now

$$M_5 = C(4 \times 144) + 2\cdot56 \times 10^3 + \sum_2^4 \Delta_i + (3\phi_2 + 2\phi_3 + \phi_4) \times 144$$

$$= -360(C + \pi_1) + 2\cdot56 \times 10^3 - 5400\lambda$$
$$= 5400\lambda + 2\cdot56 \times 10^3 - 35\cdot5 \times 360$$
$$= 12\cdot05 \times 10^3 + 2\cdot56 \times 10^3 - 12\cdot8 \times 10^3$$
$$= 1\cdot81 \times 10^3 < |M_p|.$$

Hence there is no yield at 5.

For no yield at $7-$, $8-$, $9-$ clearly $|M_i| < 2 \cdot 56 \times 10^3$. Hence all sets of inequalities will be satisfied if

$$-2 \cdot 56 < M_{7-}/10^3 < -1 \cdot 26,$$

$$-2 \cdot 56 < M_{8-}/10^3 < -1 \cdot 64,$$

$$-2 \cdot 56 < M_{9-}/10^3 < -1 \cdot 64.$$

Clearly, it can now be concluded that a possible satisfactory bending moment exists which will allow the B.M. to be the full plastic values at the required hinges for Mode I and not violate yield elsewhere, and hence that $\lambda_I = 2 \cdot 23$ represents the load factor for coincident upper and lower bounds.

Note

The type of frame discussed in this example and analysed in this manner is probably of rather rare occurrence in practice. In fact, the mode of most practical importance is Mode III, the beam mode – which here it will be noticed gives a load factor at collapse very similar to the actual factor for Mode I. In the framework proposed the columns are of lighter section than any of the beams and for a frame to be subject to sway and proportioned in this manner is unusual except under certain seismic conditions. Hence although the analysis looks a little unwieldy, it is likely to be encountered only rarely, and then probably in the symmetrical arrangement here analysed.

A point of more practical importance is that the present method of approach provides a ready method for describing bending moment distributions or displaced forms, which descriptions would be required if a computer analysis of the problem is undertaken.

Part II   The regular multi-storey frame

2.20   Introduction

Most of our discussion thus far has been concerned with principles and only occasionally have reasons been given for choosing one configuration of a structure rather than some other. We have for the most part been concerned with analysis rather than design. The subject of study here is a particular type of framework which often arises as a design need in very tall buildings, namely the structural element to resist horizontal forces arising from wind or earthquake loading conditions.

Very tall buildings frequently produce a highly repetitive structure and we shall consider the simplest possible example, the single bay plane, N storey, framework in order to obtain a feel for the characteristics of such a framework. The widths of the 'columns' of this structure in the plane of the frame are usually comparable with the span of the beams which link the columns together at regular intervals; but viewed from the point of view of the framework as a whole the column will be assumed stable and adequately described by the usual beam bending equations. The complete framework is often called a shear-wall since we anticipate that the connecting beams will cause the frame to act very much like a single plate-like element rather than a pair of columns only little influenced by the beams.

2.21   The basic problems: the equations

The two simplest problems will be discussed – the regular single bay multi-storey frame under uniformly distributed horizontal floor level loads for (*a*) the fixed (Type I) and (*b*) pinned (Type II) foundation conditions. But other more complicated foundation and loading conditions can be incorporated if desired. The typical structures of Types I and II are as shown in fig. 2.21.0.

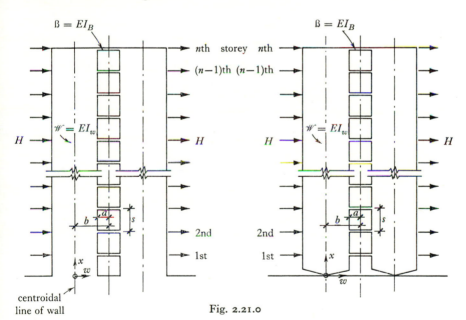

Fig. 2.21.0

Now symmetry ensures that the beam centres are points of contra-flexure and if the wall horizontal displacement is denoted by w and the value at the level of the rth beam by w_r, then the beam end-slope at level r is given by $w'_r = \mathrm{d}w_r/\mathrm{d}x$.

91

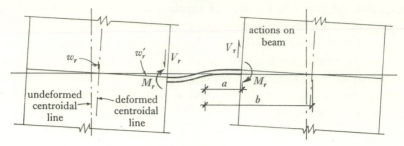

Fig. 2.21.1

Now for the beam, in terms of the column slope w_r', we have

$$M_r = \frac{6\beta w_r'}{2a} + \frac{6\beta 2(b-a)\,w_r'}{(2a)^2}$$

$$= \frac{3\beta w_r' b}{a^2} \qquad (2.21.0)$$

and $$V_r = \frac{M_r}{a}, \quad \beta = (EI)_{\text{beam}}.$$

The column experiences a moment at the centroidal axis of value:

$$\mathcal{M}_r = M_r + V_r(b-a)$$

$$= \frac{b}{a} M_r$$

$$= \frac{3\beta b^2}{a^3} w_r'. \qquad (2.21.1)$$

The column, too, although wall-like in dimensions between successive floor levels, is to be described by the beam bending equations and hence:

$$\mathcal{W} w^{\text{iv}} = p, \quad \mathcal{W} = (EI)_{\text{wall}}, \qquad (2.21.2)$$

or $$w = Ax^3 + Bx^2 + Cx + D + \text{P.I.s.} \qquad (2.21.3)$$

The particular integrals in this case are first from the external H loading and secondly from the $\mathcal{M}_r$ moments arising from bending of the beams. Hence the complete w expression becomes

$$w_r = Ax^3 + Bx^2 + Cx + D + \sum_{i=1}^{r} \frac{Hz_i^3}{3!\,\mathcal{W}} + \sum_{i=1}^{r} \frac{\mathcal{M}_i z_i^2}{2!\,\mathcal{W}} \qquad (2.21.4)$$

with $x = rs, \quad z_i = (r-i)s.$

2.21 *Basic problems: the equations*

The two cases of particular interest are the fixed and pinned foundation conditions which we shall term frames type I and II respectively. Then for frame type I, $C = D = 0$ and A, B are to be found from the $w_n'' = w_n''' = 0$ conditions at $x = ns$.

For the type II frame, $B = D = 0$ and A, C are to be found from the same moment and shear force zero conditions at $x = ns$.

Type I *frame*

At $r = n$ we have

$$0 = w_n''' = 6A + \frac{nH}{\mathcal{W}}, \\
0 = w_n'' = 6Ans + 2B + \frac{n(n-1)Hs}{2\mathcal{W}} + \sum_{i=1}^{n} \frac{\mathcal{M}_i}{\mathcal{W}}, \Bigg\} \qquad (2.21.5)$$

whence $\qquad A = -\dfrac{nH}{6\mathcal{W}} \quad$ and $\quad B = \dfrac{n(n+1)Hs}{4\mathcal{W}} - \dfrac{1}{2\mathcal{W}} \sum_{i=1}^{n} \mathcal{M}_i.$ (2.21.6)

Here we have used $\qquad \displaystyle\sum_{i=1}^{n} i = \frac{n(n+1)}{2}.$

Then from (2.21.4)

$$w_r' = 3Ax^2 + 2Bx + \sum_{i=1}^{r} \frac{Hz_i^2}{2\mathcal{W}} + \sum_{i=1}^{r} \frac{\mathcal{M}_i z_i}{\mathcal{W}}$$

$$= -\frac{nHx^2}{2\mathcal{W}} + \frac{n(n+1)Hsx}{2\mathcal{W}} + \sum_{i=1}^{r} \frac{Hz_i^2}{2\mathcal{W}} + \sum_{i=1}^{r} \frac{\mathcal{M}_i z_i}{\mathcal{W}} - \left(\sum_{i=1}^{n} \frac{\mathcal{M}_i}{\mathcal{W}}\right) x.$$

$$(2.21.7)$$

Now if we define the parameters λ, μ such that

$$\lambda \equiv \frac{a^3 \mathcal{W}}{3b^2 s\beta} \equiv \frac{1}{\mu}, \qquad (2.21.8)$$

then from (2.21.1) $\qquad \dfrac{\mathcal{M}_r . s}{\mathcal{W}} = \mu w_r', \qquad (2.21.9)$

and hence (2.21.7) are n equations in the n unknowns w_r'. These equations can be rewritten in a more convenient form as

$$(\mathbf{A} + \lambda I)\,\mathbf{w}' = \mathbf{B}, \qquad (2.21.10)$$

93

where
$$\mathbf{A} = \begin{bmatrix} 1 & 1 & 1 & 1 & 1 & 1 & \cdots & 1 \\ 1 & 2 & 2 & 2 & 2 & 2 & \cdots & 2 \\ 1 & 2 & 3 & 3 & 3 & 3 & & 3 \\ 1 & 2 & 3 & & & & & \\ 1 & 2 & 3 & & \diagdown & & & \\ \vdots & \vdots & \vdots & & & & & \\ \vdots & \vdots & \vdots & & & & & \\ 1 & 2 & 3 & & & & & n \end{bmatrix}, \quad (2.21.11)$$

$$n \times n$$

$$\mathbf{w}' = \begin{bmatrix} w_1' \\ w_2' \\ \vdots \\ w_n' \end{bmatrix} \quad (2.21.12)$$

$$n \times 1$$

and
$$\mathbf{B} = \begin{bmatrix} B_1 \\ B_2 \\ \vdots \\ B_n \end{bmatrix}; \quad B_r = \frac{Hs^2\lambda}{12\mathscr{W}}[4r^3 - 3(2n+1)r^2 + (6n(n+1)+1)r].$$

$$n \times 1 \quad\quad\quad\quad\quad (2.21.13)$$

Before discussing the solution for w_r' let us write down the corresponding expressions for the type II frame.

Type II *Frame*

As for the type I frame we have
$$0 = w_n''' = 6A + \frac{nH}{\mathscr{W}},$$

but now
$$0 = w_n'' = 6Ans + \frac{n(n-1)Hs}{2\mathscr{W}} + \sum_{i=1}^{n} \frac{\mathscr{M}_i}{\mathscr{W}},$$

with the B term absent. Hence
$$A = -\frac{nH}{6\mathscr{W}} \quad \text{and} \quad \sum_{i=1}^{n} \mathscr{M}_i = \frac{n(n+1)Hs}{2}. \quad (2.21.14)$$

This latter condition can be written alternatively as
$$\sum_{i=1}^{n} w_i' = \frac{n(n+1)Hs^2\lambda}{2\mathscr{W}}. \quad (2.21.15)$$

94

Finally then, from (2.21.4),

$$w_r' = -\frac{nH}{2\mathscr{W}}x^2 + w_0' + \sum_{i=1}^{r}\frac{Hz_i^2}{2\mathscr{W}} + \sum_{i=1}^{r}\frac{\mathscr{M}_i z_i}{\mathscr{W}}, \qquad (2.21.16)$$

where $w_0' = C =$ column slope at ground level, $r = 0$; $x = rs$ and $z_i = s(r-i)$.

There are now $n+1$ equations (2.21.15, 16) and $n+1$ unknowns $w_0', w_1', \ldots, w_r', \ldots, w_n'$.

A convenient matrix form for (2.21.16) is then

$$(\mathbf{C} + \lambda I)\mathbf{w}' - \lambda\mathbf{w}_0' = \mathbf{D}, \qquad (2.21.17)$$

where $\quad \mathbf{C} \equiv$

$$
\begin{bmatrix}
0 & 0 & \cdot & \cdot & \cdots & \cdot & 0 \\
-1 & 0 & \cdot & \cdot & \cdots & \cdot & 0 \\
-2 & -1 & 0 & \cdot & \cdots & \cdot & 0 \\
-3 & -2 & -1 & 0 & \cdots & \cdot & 0 \\
 & & & & & & 0 \\
 & & & & & & 0 \\
-(n-1) & -(n-2) & \cdots & \cdots & -2 & -1 & 0
\end{bmatrix}, \quad (2.21.18)
$$

$$n \times n$$

$$
\mathbf{w}_0' = \begin{bmatrix} 1 \\ 1 \\ 1 \\ \vdots \\ 1 \end{bmatrix} w_0', \qquad (2.21.19)
$$

$$n \times 1$$

$$
\mathbf{D} = \begin{bmatrix} D_1 \\ D_2 \\ \vdots \\ D_n \end{bmatrix} \quad \text{and} \quad D_r = \frac{Hs^2\lambda}{12\mathscr{W}}[r(r-1)(2r-1) - 6r^2]. \quad (2.21.20)
$$

$$n \times 1$$

We shall now discuss the solution of (2.21.10) and (2.21.17).

2.22   Equation solution

The problems type I and II have been reduced to the solutions of sets of n algebraic equations, (2.21.10) and (2.21.17). Difference equations suggest themselves as the vehicle for solution. Now by taking the indicated differences it can be seen that the difference equation

$$w'_{r-1} - (2+\mu)\,w'_r + w'_{r+1} = -(Hs^2/2\mathcal{W})\,[2n+1-2r] \quad (2.22.0)$$

satisfies each of (2.21.10) and (2.21.17) for $1 < r < n$.

This equation is of constant coefficient type and a solution is to be sought in the form $w'_r = g\rho^r$, where g, ρ are unknowns to be found.

Thence the value of ρ is obtained from (2.22.0) as

$$g\rho^{r-1}(1 - (2+\mu)\rho + \rho^2) = 0,$$

or

$$\rho = \frac{2+\mu \pm \sqrt{[(2+\mu)^2 - 4]}}{2}$$

$$= 1 + \frac{\mu}{2} \pm \sqrt{\left[\mu\left(1 + \frac{\mu}{4}\right)\right]}.$$

If now we define $\cosh\phi \equiv 1 + \tfrac{1}{2}\mu$ then $\sinh\phi = \sqrt{[\mu(1+\tfrac{1}{4}\mu)]}$ and $\rho_1 = \cosh\phi + \sinh\phi = e^\phi$, $\rho_2 = \cosh\phi - \sinh\phi = e^{-\phi}$.

Hence the complementary function is given by

$$w'_r = \mathcal{A}^* e^{r\phi} + \mathcal{B}^* e^{-r\phi}$$

$$= \mathcal{A}\sinh r\phi + \mathcal{B}\cosh r\phi, \quad (2.22.1)$$

where $\mathcal{A}$ and $\mathcal{B}$ are arbitrary constants.

The particular integral is sought in the form $w'_r = \mathcal{C} + \mathcal{D}r$, whence

$$\mu\mathcal{C} = \frac{Hs^2}{2\mathcal{W}}(2n+1),$$

$$\mu\mathcal{D} = -\frac{Hs^2}{\mathcal{W}}$$

Hence the general solution of (2.22.0) is

$$w'_r = \mathcal{A}\sinh r\phi + \mathcal{B}\cosh r\phi + \frac{Hs^2}{2\mu\mathcal{W}}(2n+1-2r). \quad (2.22.2)$$

The frames types I and II thus have a common general solution but must be considered separately in order to find the constants $\mathcal{A}$ and $\mathcal{B}$.

These constants are found from the conditions for $r = 1$ and n respectively, the boundary conditions.

Type I *frame*

When $r = 1$ the boundary difference condition becomes (2.21.10, 13)

$$(1 + 2\lambda) w_1' - \lambda w_2' = 2B_1 - B_2 = \frac{Hs^2\lambda}{2\mathscr{W}}(2n-1). \qquad (2.22.3)$$

Hence, using (2.22.2), this condition can be written

$$(2 + \mu)\left(\mathscr{A}\sinh\phi + \mathscr{B}\cosh\phi + \frac{Hs^2}{2\mu\mathscr{W}}(2n-1)\right)$$

$$- \left(\mathscr{A}\sinh 2\phi + \mathscr{B}\cosh 2\phi + \frac{Hs^2}{2\mu\mathscr{W}}(2n-3)\right) = \frac{Hs^2}{2\mathscr{W}}(2n-1).$$

But, by the definition of ϕ, $\cosh\phi = 1 + \tfrac{1}{2}\mu$, hence the $\mathscr{A}$ terms cancel and we obtain

$$\mathscr{B} = -\frac{(2n+1)Hs^2}{2\mu\mathscr{W}}. \qquad (2.22.4)$$

The condition at $r = n$ is evidently

$$(1 + \lambda) w_n' - \lambda w_{n-1}' = \frac{Hs^2\lambda}{2\mathscr{W}}, \qquad (2.22.5)$$

or

$$(1 + \mu)\left(\mathscr{A}\sinh n\phi + \mathscr{B}\cosh n\phi + \frac{Hs^2}{2\mu\mathscr{W}}\right)$$

$$- \left(\mathscr{A}\sinh(n-1)\phi + \mathscr{B}\cosh(n-1)\phi + \frac{3Hs^2}{2\mu\mathscr{W}}\right) = \frac{Hs^2}{2\mathscr{W}}. \qquad (2.22.6)$$

There are two frequently occurring functions of ϕ which can usefully be given definitions. Thus we shall define P_n and Q_n to be

$$P_n \equiv \frac{\mu\sinh n\phi}{\sinh\phi}; \quad Q_n \equiv \cosh n\phi + \frac{\mu}{2}\frac{\sinh n\phi}{\sinh\phi}$$

$$= \cosh n\phi + \frac{P_n}{2}. \qquad (2.22.7)$$

In addition, the following identities will be used as necessary

$$\left.\begin{array}{l}(1 + \mu)\cosh n\phi - \cosh(n-1)\phi = \tfrac{1}{2}\mu\cosh n\phi + \sinh n\phi \sinh\phi, \\[2mm] (1 + \mu)\sinh n\phi - \sinh(n-1)\phi = Q_n\sinh\phi\end{array}\right\} \qquad (2.22.8)$$

and $\qquad Q_{n-r} = Q_n\cosh r\phi - \left(\sinh n\phi + \frac{\mu\cosh n\phi}{2\sinh\phi}\right)\sinh r\phi.$

Equation (2.22.6) can now be simplified to

$$\mathscr{A}Q_n \sinh\phi + \mathscr{B}(Q_n \cosh r\phi - Q_{n-r})\frac{\sinh\phi}{\sinh r\phi}$$

$$= \frac{Hs^2}{2\mu\mathscr{W}}(\mu-(1+\mu)+3) = \frac{Hs^2}{\mu\mathscr{W}},$$

or $\quad Q_n(\mathscr{A}\sinh r\phi + \mathscr{B}\cosh r\phi) = \mathscr{B}.Q_{n-r} + \frac{Hs^2.P_r}{\mu^2\mathscr{W}}$

$$= \frac{Hs^2}{2\mu\mathscr{W}}(-(2n+1)Q_{n-r}+2\lambda P_r).$$

Hence, using this expression in (2.22.2) gives

$$w'_r = \frac{Hs^2}{2\mu\mathscr{W}}\frac{1}{Q_n}((2n+1-2r)Q_n - (2n+1)Q_{n-r} + 2\lambda.P_r). \quad (2.22.9)$$

This expression (2.22.9) is the basic result for the type I frame.

Type II frame

In this case when $r=1$ the boundary difference condition becomes

$$(1+\lambda)w'_1 - \lambda w'_2 = \frac{Hs^2\lambda}{2\mathscr{W}}(3n-1), \quad (2.22.10)$$

or

$$(1+\mu)\left(\mathscr{A}\sinh\phi + \mathscr{B}\cosh\phi + \frac{Hs^2}{2\mu\mathscr{W}}(2n-1)\right)$$

$$-\left(\mathscr{A}\sinh 2\phi + \mathscr{B}\cosh 2\phi + \frac{Hs^2}{2\mu\mathscr{W}}(2n-3)\right) = \frac{Hs^2}{2\mathscr{W}}(3n-1).$$

$$(2.22.11)$$

Note that the w'_0 value does not enter this expression. In contrast to the type I frame first boundary condition, no simple relation results for $\mathscr{B}$.

The second boundary difference condition is now

$$\sum_{i=1}^{n} w'_i - (1+\lambda).w'_n + \lambda.w'_{n-1} = \frac{Hs^2\lambda}{2\mathscr{W}}(n(2n-1)-(n-1)^2). \quad (2.22.12)$$

Again w'_0 is absent.

The first term can be eliminated by use of (2.21.15) and, simplifying, we obtain

$$-(1+\mu)w'_n + w'_{n-1} = -\frac{Hs^2}{2\mathscr{W}},$$

or

$$(1 + \mu) \left(\mathscr{A} \sinh n\phi + \mathscr{B} \cosh n\phi + \frac{Hs^2}{2\mu\mathscr{W}} \right)$$
$$- \left(\mathscr{A} \sinh (n-1)\phi + \mathscr{B} \cosh (n-1)\phi + \frac{3Hs^2}{2\mu\mathscr{W}} \right) = \frac{Hs^2}{2\mathscr{W}}. \quad (2.22.13)$$

Use of $(2.22.8)_2$ then gives

$$\mathscr{A} . Q_n \sinh n\phi + \mathscr{B}((1 + \mu) \cosh n\phi - \cosh (n-1)\phi) = \frac{Hs^2}{\mu\mathscr{W}}, \quad (2.22.14)$$

and (2.22.11) simplifies to

$$\mathscr{A} \sinh \phi + \mathscr{B}(\cosh \phi - 1) = \frac{Hs^2}{2\mu\mathscr{W}} (2 - \mu n). \quad (2.22.15)$$

Used with (2.22.14) this reduces to

$$\mathscr{A} \sinh r\phi + \mathscr{B} \cosh r\phi = \mathscr{B} \frac{Q_{n-r}}{Q_n} + \frac{Hs^2}{\mu^2\mathscr{W}} \frac{P_r}{Q_n}. \quad (2.22.16)$$

From (2.22.15, 16) $\mathscr{B}$ can be found to be

$$\mathscr{B} = \frac{4 + \mu}{8} \frac{2 + Q_n(\mu n - 2)}{\sinh \phi \sinh n\phi} \frac{Hs^2}{\mu\mathscr{W}}.$$

Let us define

$$R(n, \mu) \equiv \frac{2\mu\mathscr{W}\mathscr{B}}{Hs^2} = \frac{4 + \mu}{4} \frac{2 + Q_n(\mu n - 2)}{\sinh \phi . \sinh n\phi}, \quad (2.22.17)$$

then the expression for w'_r in this case, after substituting (2.22.17) into (2.22.2), becomes

$$w'_r = \frac{Hs^2}{2\mu\mathscr{W}} \frac{1}{Q_n} ((2n + 1 - 2r) Q_n + R(n, \mu) Q_{n-r} + 2\lambda P_r). \quad (2.22.18)$$

This is the basic result for the type II frame.

Hence the expressions for w'_r for the types I and II frames are identical except for the coefficient of the Q_{n-r} term.

With the expressions (2.22.9) and (2.22.18) to hand, all quantities of interest can be easily computed. The quantities of particular interest are the axial forces in the columns at ground level since they are a sensitive measure of the interaction between the two columns; and the maximum deflexions. In the case of the type II frame the vertical column loads at ground level can be found from overall statics of the frame. For the type I frame it is convenient to evaluate the moments in the columns at ground level and then find the vertical forces from statics.

Now for the type I frame

$$\mathcal{M}_0 = \mathcal{W} w_0'' = 2\mathcal{B}\mathcal{W}$$

$$= \frac{n(n+1)}{2} Hs - \sum_{i=1}^{n} \mathcal{M}_i, \qquad (2.22.19)$$

after use of (2.21.6).

From the definitions of ϕ, P_r, Q_r it can be seen that

$$P_r = Q_r - Q_{r-1}, \quad Q_r = \lambda(P_{r+1} - P_r),$$

and hence

$$\left.\begin{array}{c} \displaystyle\sum_{i=1}^{n} P_i = Q_n - 1, \\[2ex] \displaystyle\sum_{i=1}^{n} Q_i = \lambda P_{n+1} - 1, \\[2ex] \displaystyle\sum_{i=1}^{n} Q_{n-i} = \lambda P_n. \end{array}\right\} \qquad (2.22.20)$$

Hence

$$\sum_{i=1}^{n} \mathcal{M}_i = \frac{Hs}{2Q_n} \left((n(2n+1) - n(n+1)) Q_n - (2n+1)\lambda P_n + 2\lambda(Q_n - 1) \right)$$

$$= \frac{Hs}{2Q_n} \left((n^2 + 2\lambda) Q_n - \lambda(2n+1) P_n - 2\lambda \right). \qquad (2.22.21)$$

Now let $\mathcal{V}$ be the vertical force acting through the column centroid and caused by the lateral loading, then by overall statics

$$2\mathcal{V}.b = n(n+1) Hs - 2\mathcal{M}_0. \qquad (2.22.22)$$

For the type I frame $\mathcal{M}_0$ is given by (2.22.19) and the expression (2.22.22) can be further simplified to

$$\mathcal{V}.b = \sum_{i=1}^{n} \mathcal{M}_i = \frac{Hs}{2Q_n} \left((n^2 + 2\lambda) Q_n - \lambda(2n+1) P_n - 2\lambda \right). \quad (2.22.23)$$

For the type II frame $\mathcal{M}_0 = 0$ and $\mathcal{V}$ can be found directly from (2.22.22). To complete the information required from the solutions (2.22.9) and (2.22.18) we shall evaluate the roof level displacements.

2.23   The displacements

In order to compute the displacements from (2.21.4) various summations must be made. Now from (2.21.4) and (2.21.6), for the type I frame,

$$w_n = \mathcal{A}(ns)^3 + \mathcal{B}(ns)^2 + \sum_{i=1}^{n} \frac{Hz_i^3}{3!\,\mathcal{W}} + \sum_{i=1}^{n} \frac{\mathcal{M}_i z_i^2}{2!\,\mathcal{W}}, \qquad (2.23.0)$$

with

$$\mathscr{A} = -\frac{Hn}{6\mathscr{W}}, \qquad \mathscr{B} = \frac{n(n+1)\,Hs}{4\mathscr{W}} - \frac{1}{2\mathscr{W}}\sum_{i=1}^{n}\mathscr{M}_i$$

and the $\mathscr{M}_i$ given by (2.21.1), and (2.22.9).

Evidently the following summations are required:

$$
\left.
\begin{aligned}
\sum_{r=1}^{n} rP_r &= \sum_{r=1}^{n} r(Q_r - Q_{r-1}) = \sum_{1}^{n} rQ_r - \sum_{0}^{n-1}(r+1)Q_r \\
&= nQ_n - \sum_{1}^{n-1} Q_r = nQ_n - \lambda P_n. \\[4pt]
\sum_{r=1}^{n} rQ_r &= \lambda(nP_{n+1} - Q_n + 1). \\[4pt]
\sum_{r=1}^{n} rP_{n-r} &= \sum_{r=0}^{n-1}(n-r)P_r = \lambda P_n - n. \\[4pt]
\sum_{r=1}^{n} rQ_{n-r} &= \lambda(Q_n - 1). \\[4pt]
\sum_{r=1}^{n} r^2 P_r &= \sum_{r=1}^{n} r^2(Q_r - Q_{r-1}) = \sum_{r=1}^{n} r^2 Q_r - \sum_{r=1}^{n+1}(r+1)^2 Q_r \\
&= (n^2 + 2\lambda)Q_n - (2n+1)\lambda P_n - 2\lambda. \\[4pt]
\sum_{r=1}^{n} r^2 Q_r &= (n^2 - 2n\lambda + \lambda)Q_n + \lambda(n^2 + 2\lambda)P_n - \lambda. \\[4pt]
\sum_{r=1}^{n} r^2 P_{n-r} &= 2\lambda Q_n - \lambda P_n - (n^2 + 2\lambda). \\[4pt]
\sum_{r=1}^{n} r^2 Q_{n-r} &= \lambda Q_n + 2\lambda^2 P_n - (2n+1)\lambda.
\end{aligned}
\right\} \qquad (2.23.1)
$$

Hence for the type I frame by easy steps it can be shown that

$$
w_n = \frac{Hs^3}{2\mathscr{W}}\left(\frac{n^2(n+1)(3n+1)}{12} - \left\{\frac{n(3n^3 + 4n^2 - 1)}{12} - \lambda(n^2 - \tfrac{1}{2} + 2\lambda)\right.\right.
$$

$$
\left.\left. + 2\lambda^2(n+1)\frac{P_n}{Q_n} - \lambda(n + \tfrac{1}{2} - 2\lambda)\frac{1}{Q_n}\right\}\right). \qquad (2.23.2)
$$

The term in the brace is never negative and tends to zero as $\mu \to 0$. It represents the stiffening effect of the beams. The first term in the outer bracket is the deflexion of the bare cantilever column in the case $\mu \to 0$, namely no beam moments.

The companion result for the type II frame can be found from (2.21.0) using now (2.22.18) to give the column slopes. The result is

$$w_n = \frac{Hs^3}{6\mathscr{W}}\left(-n^4+\frac{3n}{\mu Q_n}\{(2n+\mu n-1)Q_n+R_nQ_{n-1}+2\}+\frac{n^2(n-1)^2}{4}\right.$$

$$+\frac{3}{Q_n}\left\{\frac{n(n-1)(3n^2-n-1)Q_n}{6}+((1-2n)Q_n+(n^2+2\lambda)P_n-1)\lambda R_n\right.$$

$$\left.\left.+2\lambda(2\lambda Q_n-\lambda P_n-(n^2+2\lambda))\right\}\right), \tag{2.23.3}$$

with R_n given by (2.22.17).

Here there is no such easy interpretation as for (2.23.2). The displacement w_n is positive and tends to infinity as $\mu \to 0$ since then the framework tends towards a mechanism.

The rotation of the column at ground level, w_0', is given by (2.21.16) for $r = 1$ and is

$$w_0' = w_1'+\frac{Hs^2n}{2\mathscr{W}}$$

$$= \frac{Hs^2}{2\mu\mathscr{W}Q_n}((2n+\mu.n-1)Q_n+R_n.Q_{n-1}+2). \tag{2.23.4}$$

This quantity too is unbounded for $\mu \to 0$.

Exercise

As a check on the correctness of the results (2.23.2) and (2.23.3), solve the single storey fixed and pinned base conventional portal problems for a uniform section and span twice the height. Show that in the present notation then $n = 1$, $\mu = 3$, $P_1 = 3$, $Q_1 = 4$, $Q_0 = 1$ and hence that for the type I frame

$$w_1' = \frac{Hs^2}{8\mathscr{W}}, \quad w_1 = \frac{7Hs^3}{48\mathscr{W}},$$

and for the type II frame

$$w_1' = \frac{Hs^2}{3\mathscr{W}}, \quad w_0' = \frac{5Hs^2}{6\mathscr{W}}, \quad w_1 = \frac{2Hs^3}{3\mathscr{W}}.$$

Note

2.10 Much of the notation used in this discussion of shear walls and the results, especially (2.22.9) and (2.23.2), are due to J. Blanchard. I read and made notes on his unpublished analysis and induction proof, in 1958. But in the present treatment I have preferred to give a different development and a solution which avoids induction arguments while at the same time I have added a parallel analysis for the pin-base frame. But then as now I value the Blanchard analysis, which has provided most of my insight into this problem.

2.24 Discrete and continuous formulations

All the foregoing discussion is in terms of discrete beams with their usual characteristics. In the discussion of composite beams in chapter 4, although the shear connexion is achieved by use of individual shear connectors, the model adopted for the description of connexion is a *continuous* connexion parameter, λ.

The same procedure can be adopted for the shear wall. In fact it will be seen that, with suitable identification of parameters, the analysis of this chapter for the shear wall can be taken over and used for a discrete theory of shear connexion and likewise the continuous theory for the shear connexion provides an alternative formulation for the shear wall problem. The literature on this subject is not large but the significant papers in this area (see references) usually use a continuous formulation for the shear wall problem. The present analysis provides a direct link with the customary frame parameters of beam stiffness, etc. and might be of use in choosing suitable parameters when adopting a continuous formulation.

Clearly the two formulations should merge one with another for large n. The formal similarities are however evident for all values of n. Thus we shall see that the analogous composite beam continuous formulation is expressible as cosh and sinh of a certain *continuous variable*. The discrete problem as we have seen is expressible in terms of cosh and sinh of a certain *discrete variable* (for example w'_r as given by (2.22.2)). The correspondence is $\beta s = \sqrt{\mu}$, β being defined in (4.7.5).

2.25 An example: exercises

In the sections 2.20–2.23 we have established various results relating to the regular single bay multi-storey frame, better described as a shear wall structure. In table 2.25.0 (appendix) are collected together, for a range of values of the parameters, a series of results which should be useful when applying the analysis of this chapter. Primarily the table gives values for P_n, Q_n, R_n (2.22.7, 2.22.17) but also tabulated is the value S, the percentage relief of moment at the base of the wall, that is the percent fraction by which the pure cantilever moment, which would develop if $\mu = 0$, is reduced in the actual shear wall. This value of course applies only to the type I frame.

Thus, from (2.22.23):

$$S \equiv \frac{100}{n(n+1)} \frac{1}{Q_n} (Q_n(n^2 + 2\lambda) - P_n \lambda(2n+1) - 2\lambda). \qquad (2.25.0)$$

The final two columns in the table give values of the roof-level deflexion for the types I and II frames, respectively, each divided by $\dfrac{nH(ns)^3}{8\mathcal{W}}$.

Then $w_I \equiv$ roof-level displacement of type I frame

$$= (\text{coefficient from table}) \times \frac{nH(ns)^3}{8\mathcal{W}}, \qquad (2.25.1)$$

and similarly for the type II frame.

The pure cantilever displacement, that is the value of w_I for $\mu \to 0$, is given by

$$(w_I)_{\mu=0} = \left(1 + \frac{1}{n}\right)\left(1 + \frac{1}{3n}\right)\frac{nH(ns)^3}{8\mathcal{W}}, \qquad (2.25.2)$$

and (2.25.2) is clearly an upper limit on the value of w_I for given n and all μ, since the connecting beams always act to reduce the displacements.

As an example, consider a shear wall of uniform thickness (t), 17 metres wide and 75 metres high. Assume that the foundations permit rigidity of base connexion comparable to the type I frame and that the wall is pierced with a regular series of rectangular openings, spaced at 3 m vertical intervals, 3 m wide and on the wall centre-line. An opening depth of 2·5 m leaves an 0·5 m deep beam spanning across between the 7 m wide walls as shown in fig. 2.25.0. The actual wall thickness is not required for the present.

Now here $ß \propto (\frac{1}{2})^3$, $w \propto 7^3$, $a = 1·5$ m, $b = 5$ m, $s = 3$ m, $n = 25$. Hence, in (2.21.8)

$$\mu = \frac{3 \times 5^2 \times 3 \times 0·5^3}{1·5^3 \times 7^3} = 0·0243. \qquad (2.25.3)$$

From table 2.25.0 we see that

$$S \doteqdot 62. \qquad (2.25.4)$$

2.25 An example

This implies that if the beams had been neglected, each 7 m wide wall would act as a cantilever and the base moment would be given by

$$(\mathscr{M}_0)_{\mu=0} = \frac{n(n+1)}{2} Hs. \qquad (2.25.5)$$

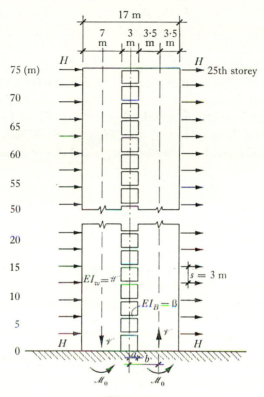

Fig. 2.25.0

But because of the beams this moment is reduced by 62 % and the actual wall moment at foundation level will have a value of

$$\mathscr{M}_0 = 0.38 \frac{n(n+1)}{2} Hs. \qquad (2.25.6)$$

The remaining moment is supported by vertical forces, $\mathscr{V}$, as shown on fig. 2.25.0, and given from (2.22.23, 2.25.1) as

$$\mathscr{V} = \frac{n(n+1) SHs}{200b}. \qquad (2.25.7)$$

This development of vertical forces is very beneficial in so far that resultant wall tensions can often be avoided or at least minimized. They may however introduce stability problems in certain circumstances.

In addition to wall moments and forces the beam moments may be evaluated from (2.22.9, 18). In general the maximum beam moment occurs around mid-height in the shear wall. Deflexions can also be found from table 2.25.0. The interested reader should set about exploring this table.

Table 2.25.1 (appendix) gives a tabulation for P_n, Q_n, R_n, etc. in the case when μ is of the order unity and n is small. Such large values for μ are characteristic of frameworks rather than shear walls. This table can be used in conjunction with the concluding sections of Part I of this chapter to investigate regular single bay frames of conventional beam and column construction, of a small number of storeys and under lateral loading.

exercises

The regular pattern of the matrix of coefficients for the type I frame (2.21.11) suggests it might be possible to find the inverse for this matrix and the coefficient matrix in (2.21.10). We leave these tasks as exercises for the reader. We begin the exercises with a study of some simpler but related matrices and their inverses.

(1) Given

$$
\mathscr{A} = \begin{bmatrix}
1 & 1 & 1 & 1 & 1 & \cdots & 1 \\
1 & 2 & 2 & 2 & 2 & \cdots & 2 \\
1 & 2 & 3 & 3 & 3 & \cdots & 3 \\
1 & 2 & 3 & & & & \\
\vdots & \vdots & \vdots & & & & \\
\vdots & \vdots & \vdots & & & & \\
1 & 2 & 3 & & & & n
\end{bmatrix},
$$

an $n \times n$ symmetric matrix, show that $\mathrm{Det}\,\mathscr{A}$ is unity and hence that $\mathscr{A}^{-1}$ is given by

2.25 *An example*

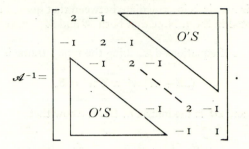

$$\mathcal{A}^{-1} = \begin{bmatrix} 2 & -1 & & & & & \\ -1 & 2 & -1 & & O'S & & \\ & -1 & 2 & -1 & & & \\ & & -1 & 2 & -1 & & \\ O'S & & & \ddots & -1 & 2 & -1 \\ & & & & & -1 & 1 \end{bmatrix}.$$

Note the 1 rather than a 2 in the trailing bottom right position.

(2) If the 1 in the (n, n) position in $\mathcal{A}^{-1}$ of Ex. 1 is replaced by a 2, then the simple $\mathcal{A}$ matrix is changed quite radically and is included in the following family of $\mathcal{B}$ matrices.

Given

$$\mathcal{B} = \begin{bmatrix} P & -1 & & & & & \\ -1 & P & -1 & & O'S & & \\ & -1 & P & -1 & & & \\ & & -1 & P & -1 & & \\ O'S & & & \ddots & -1 & P & -1 \\ & & & & & -1 & P \end{bmatrix},$$

then

$$\mathcal{B}^{-1} = \frac{1}{\mathcal{Q}_n} \begin{bmatrix} \mathcal{Q}_0 \cdot \mathcal{Q}_{n-1} & \cdots & \mathcal{Q}_0 \cdot \mathcal{Q}_{n-r} & \cdots & \mathcal{Q}_0 \cdot \mathcal{Q}_0 \\ \vdots & \ddots & \vdots & & \vdots \\ \mathcal{Q}_0 \cdot \mathcal{Q}_{n-r} & \cdots & \mathcal{Q}_{r-1} \cdot \mathcal{Q}_{n-r} & \cdots & \mathcal{Q}_{r-1} \cdot \mathcal{Q}_0 \\ \vdots & & \vdots & \ddots & \vdots \\ \mathcal{Q}_0 \cdot \mathcal{Q}_0 & \cdots & \mathcal{Q}_{r-1} \cdot \mathcal{Q}_0 & \cdots & \mathcal{Q}_{n-1} \cdot \mathcal{Q}_0 \end{bmatrix} \begin{matrix} \text{1st} \\ \\ \text{rth} \\ \\ \text{nth} \end{matrix}$$

where
$$\mathcal{Q}_n = \frac{\sinh(n+1)\phi}{\sinh \phi}; \quad \cosh \phi \equiv \frac{P}{2}.$$

(See Bickley and McNamee, 2.26.)

In a sense this matrix is intermediate between that quoted in Ex. 1 and that about to be quoted in Ex. 3. Note that both $\mathscr{B}$ and $\mathscr{B}^{-1}$ have symmetry about *both* diagonals.

(3) The system of equations describing the type I frame is

$$(\mathscr{A} + \lambda I)\,\mathbf{w}' \equiv \mathscr{C}\mathbf{w}' = \mathbf{B},$$

from (2.21.10), and $\mathscr{A}$ is the matrix of Ex. 1. Show that

$$\lambda\mathscr{C}^{-1} = I - \frac{1}{Q_n}
\begin{bmatrix}
P_1 \cdot Q_{n-1} & \cdots & P_1 \cdot Q_{n-r} & \cdots & P_1 \cdot Q_0 \\
\vdots & & \vdots & & \vdots \\
\vdots & & \vdots & & \vdots \\
\vdots & & \vdots & & \vdots \\
P_1 \cdot Q_{n-r} & \cdots & P_r \cdot Q_{n-r} & \cdots & P_r \cdot Q_0 \\
\vdots & & \vdots & & \vdots \\
\vdots & & \vdots & & \vdots \\
P_1 \cdot Q_0 & \cdots & P_r \cdot Q_0 & \cdots & P_n \cdot Q_0
\end{bmatrix},$$

where

$$P_n = \mu\,\frac{\sinh n\phi}{\sinh \phi}, \quad Q_n = \cosh n\phi + \frac{P_n}{2},$$

$$\cosh \phi \equiv 1 + \frac{\mu}{2}, \quad \lambda\mu = 1.$$

This is an unpublished result due to J. Blanchard. Here the structure of the inverse is a little more complicated but note the structure of the $\mathscr{A}$ matrix, with the repetition to the right and below a given diagonal element, carried over, in part at least, to the inverse.

(4) The problem formulated in 2.21 is based on the assumption of inextensible walls and beams. That is to say, the bending neutral axis of the member is assumed to be unstrained by any resultant axial forces present. A more realistic assumption would be to assume a proportional response between neutral axis deformation and axial force. Below we give brief details of an analysis for the case of an extensible wall, and leave it to the reader to explore the problem further. The notation is as in 2.21–5.

Thus, if the neutral axis axial displacement is denoted by u_r at the rth floor level, assumed positive tensile, then the beam moments are

$$M_r = \frac{3\beta b}{a^2}\,(w'_r - u_r/b),$$

and the wall moments are

$$\mathcal{M}_r = \frac{3\beta b^2}{a^3}(w'_r - u_r/b).$$

The resultant axial wall forces will be denoted by $\mathcal{V}_r$ and measured at r^+. Then, in the absence of wall self weight, vertical equilibrium is described by

$$\mathcal{V}'_r = 0,$$

and the force–displacement $(\mathcal{V}_r, u_r)$ relation is assumed to be the linear elastic relation

$$\mathcal{V}_r = AEu'_r.$$

Here A is the wall area.

Hence
$$u''_r = 0,$$

or
$$u_r = \alpha x + \beta + \sum_1^r \frac{v_i z_i}{AE};$$

the v_i are the vertical beam shears,

$$v_i = \frac{|M_i|}{a} = -\frac{3\beta b}{a^2}(w'_r - u_r/b),$$

in the tensile wall, and the z_i's indicate discontinuous behaviour. The boundary conditions on u_r are $u_0 = 0$, $u'_n = 0$, hence

$$\beta = 0, \quad AE\alpha = -\sum_1^n v_i.$$

If we now refer back to (2.21.7–13), it is easy to see that (2.21.10), modified to

$$(A + \lambda I)\,w' = B + (A)\,u/b \tag{a}$$

describes the wall deformation and is here coupled with

$$(A + \theta I)\,u/b = (A)\,w', \tag{b}$$

where
$$\theta = \frac{AEa^3}{3\beta s}.$$

These two $(n \times n)$ matrix equations, on addition, give

$$\lambda w' + \frac{\theta}{b}u = B, \tag{c}$$

which $(n \times 1)$ vector equation can be used to eliminate either u or w' from (a) or (b).

Thus, (c) used in (b) to eliminate u yields

$$\left(A + \frac{\lambda}{\alpha}I\right)w' = \frac{1}{\alpha}\left(I + \frac{A}{\theta}\right)B, \tag{d}$$

where $\alpha \equiv 1 + \lambda/\theta \geqslant 1$; while eliminating $\mathbf{w}'$ gives

$$\left(\mathbf{A} + \frac{\lambda}{\alpha}I\right)\frac{\mathbf{u}}{b} = \frac{(\mathbf{A})}{\alpha\theta}\mathbf{B}. \tag{e}$$

These latter two matrix equations suggest a new variable $\mathbf{w}' - \mathbf{u}/b$ and then

$$\left(\mathbf{A} + \frac{\lambda}{\alpha}I\right)\left(\mathbf{w}' - \frac{\mathbf{u}}{b}\right) = \frac{1}{\alpha}\mathbf{B}. \tag{f}$$

(f) is the significant equation since this is completely analogous to (2.21.10) and the results flowing there from can all be translated into the present context by a simple exchange of λ there by λ/α and H there by H/α.

Hence the tables 2.25.0 and 2.25.1 can be used immediately to investigate shear wall and frame structures with extensible walls.

For example, the shear wall analysed in 2.25, now reanalysed as a compressible wall system gives

$$\alpha = 1 + \frac{\lambda}{\theta} = 1 + \frac{\mathscr{W}}{b^2 EA} = 1 + \frac{7^2}{12 \times 5^2} = 1.163.$$

Or, modified, $\qquad \mu = 1.163 \times 0.0243 = 0.0283.$

From table 2.25.0, by interpolation,

$$S \doteqdot 63,$$

$$\mathscr{M}_0 = 0.37 \frac{n(n+1)}{12} \frac{Hs}{1.16}$$

and

$$\mathscr{V} = \frac{n(n+1) \times 63 \times Hs}{200 \times b \times 1.16},$$

namely, $\mathscr{V}$ here is $63/62 \times 1.16 = 0.876$ of the value obtained by ignoring compressibility. The effect of wall compressibility is to lower the estimates for interaction moments and shears obtained from the inextensional theory.

The role of α in this analysis has been discussed by a former colleague, R. Cooper, in an unpublished article in 1960.

2.26   References

Asplund, S. O. *Structural Mechanics: Classical and Matrix Methods*, Prentice-Hall, New Jersey (1966). A modern text.

Baker, Sir John. *The Steel Skeleton*, vol. 1 (1954), vol. 11 (1956), with M. R. Horne and J. Heymann, Cambridge University Press, London. The first volume deals with elastic behaviour and the background to the developments in plastic theory. Plastic theory is dealt with in volume 11.

2.26 *References*

Bickley, W. G. and McNamee, J. *Phil. Trans. Roy. Soc.* (A) (1960), **252**, 69.

Coull, A. and Stafford-Smith, B. (editors). *Tall Buildings*, A symposium, Pergamon, London (1967). A general reference but the reader can find more specific works listed in the various papers in this collection.

Cross, H. and Morgan, N. D. *Continuous Frames of Reinforced Concrete*, Wiley, London (1932). Moment distribution, column analogy and arches. A pioneer text.

Grinter, L. E. *Theory of Modern Steel Structures*, vol. II, Macmillan, New York (1949). An American text, now largely out of date, but a widely used book twenty years ago. A good example of a text from that period.

Hall, A. S. and Woodhead, R. W. *Frame Analysis*, Wiley, London (1965), 2nd edition. Matrices, flexibility and stiffness methods. Examples with answers, a good modern text.

Heyman, J. On the estimation of deflexions in elastic–plastic framed structures *Proc. I.C.E.* (1960), **19**, 39.

Hodge, P. G. *Plastic Theory of Structures*, McGraw-Hill, New York (1959). A popular modern treatment of plastic structures. Problems without answers.

Neal, B. G. *The Plastic Methods of Structural Analysis*, Chapman and Hall, London (1963), 2nd edition. A standard text with the subject developed from work principles.

Parcel, J. L. and Maney, G. H. *An Elementary Treatise on Statically Indeterminate Stresses*, Wiley, London (1936), 2nd edition. Useful historical bibliography and notes.

Pippard, A. J. S. and Baker, Sir John. *The Analysis of Engineering Structures*, Arnold, London (1968), 4th edition. A widely used text, first published in 1936. Emphasis on elastic methods.

3

THE COLUMN

3.0   Introduction

The member to be studied in this chapter is the *thrust* (compression) resisting member, the *column*, usually met with as a vertical member in a building frame. The behaviour of the column can be contrasted with that of the beam, typically a horizontal member.

In the case of the column, the describing or G.D.E. is only partially integrable in quadratures, where for the beam the G.D.E. is fully integrable in quadratures. This has consequences for the discontinuous particular integral calculations. Again, the range of physical phenomena associated with column behaviour is larger than for beams in so far that in addition to strength and stiffness problems, columns also exhibit instability characteristics. Instability happenings centre around the sudden and radical changes in behaviour of a member in which *compressive* forces play a leading role. The changes take place at some critical value of the load, called variously the buckling, critical or instability load. For the column, the compressive force of interest is the axial force.

An important feature of the analysis we are about to present is the discussion of forces and displacements together at the outset. This is a logical consequence of our earlier decision for beam analysis and serves to reinforce that decision, since by treating the problems in this way the close connexions between beam and column problems, the order of the equations, the treatment of boundary conditions and so on, can be seen to be similar if not identical.

The point was made in discussing the beam equilibrium equations that the displacements and slopes were assumed to be sufficiently small for the equations written in terms of the initial shape of the member to be sufficiently accurate. The difference, the only difference, between a beam and a column from the point of view of the describing equations is that for the column, *equilibrium is written down for the deformed configuration of the element*. From this one difference, very considerable differences in behaviour result. In particular, instability phenomena now appear.

For beams we have assumed at least one axis of symmetry for the beam section in order to ensure a common plane for loads and displacements, in addition to ensuring a common practical requirement. The most important column elements in practice have two axes of symmetry of the section. But section symmetry is not essential to the further discussion of instability. In all cases of instability however the section properties, which will occur as the combination EI, must be chosen to be the (EI) minimum for the section. Independently of symmetry, this value exists about a principal axis which will be an axis of symmetry, if there is any symmetry.

Before commencing our own discussion of column behaviour we would impress upon the reader our implicit requirement that he should consult other discussions of column behaviour, particularly those with a bias for the physical behaviour since we are well aware of gaps in our own discussion and we shall assume that the reader will fill these in from other sources. As suggestions for collateral reading see especially Drucker, Bleich and Galambos, (3.16).

The subject of column behaviour is a wide field and capable of various emphases. Our discussion follows the historical sequence and treats mainly of elastic behaviour. The methods which we shall employ however are somewhat novel. We shall discuss in some detail the properties of the discontinuous solutions appropriate to the present problem and this will form yet another link with our earlier discussion of beams and frameworks.

3.1 The fundamental equations and solutions

A typical column element is shown in fig. 3.1.0. By resolving in two directions and taking moments for the element in its *final* position the following equilibrium equations are obtained:

$$dP = 0, \quad \frac{dF}{dx} = p, \quad \frac{dM}{dx} + P\frac{dw}{dx} = F. \tag{3.1.0}$$

The new term over those obtained for the similar situation for the beam is the term $P \, . \, dw/dx$ which arises essentially as a force $\times$ displacement, a deformation moment, term. All the other terms are of the form force $\times$ dimension of the structure. The equations (3.1.0) are three equations in four unknowns, P, F, M, w. The set is therefore insoluble. The displacements are involved at the outset and hence there is no class of problems corresponding to the determinate beam problem; but the above equations are of course material property independent.

The set (3.1.0) must be supplemented with the material property equations. For our discussion the elastic material property will be assumed, but others are possible. This simple material property assumption is important historically and practically and produces a theory which possesses all the essential features associated with real column behaviour. The material property equation we shall assume is exactly as for the beam, namely

$$M = EI\frac{\mathrm{d}^2w}{\mathrm{d}x^2} = EIw''. \tag{3.1.1}$$

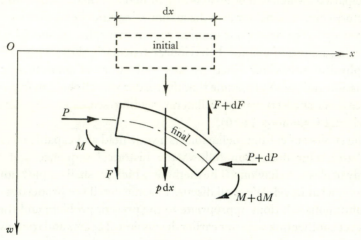

Fig. 3.1.0

If F and M are eliminated from (3.1.0, 1) we obtain an equation for w, the transverse displacement, namely

$$(EIw'')'' + Pw'' = p. \tag{3.1.2}$$

In all cases with which we shall deal $EI = $ const., namely the member is prismatic, and, with $P = $ const. from (3.1.0), we can write (3.1.2) as:

$$w^{\mathrm{iv}} + \alpha^2 w'' = p/EI, \quad \alpha^2 = P/EI = \text{const.} \tag{3.1.3}$$

This equation is the fundamental equation of the present study. A solution can now be sought in the form $w = Ae^{\lambda x}$ when, for the complementary function,

$$\lambda^4 + \alpha^2\lambda^2 = 0, \tag{3.1.4}$$

or

$$\lambda = 0, 0, i\alpha, -i\alpha. \tag{3.1.5}$$

A convenient form of the solution is therefore

$$w = A\sin\alpha x + B\cos\alpha x + Cx + D + \text{P.I.} \tag{3.1.6}$$

If p is a constant u.d.l. then the P.I. is $px^2/2P$.

3.1 *Fundamental equations and solutions*

As with the beam, so too for the column, there are *four* arbitrary constants which must be found from *two* conditions at each of the two ends of the column length. These conditions are again referred to as the *boundary conditions*. Typical examples of these conditions are:

$$
\left.\begin{array}{lll}
(\text{1}) \ \text{Pinned end} & w = w'' = \text{o}, & \\
(\text{2}) \ \text{Fixed end} & w = w' = \text{o}, & \\
(\text{3}) \ \text{Free end} & w'' = w''' + \alpha^2 w' = \text{o}, & \\
(\text{4}) \ \text{Axis of symmetry} & w' = w''' + \alpha^2 w' = \text{o}, & \\
(\text{5}) \ \text{Axis of antisymmetry} & w = w'' = \text{o}. &
\end{array}\right\} \qquad (3.1.7)
$$

All these conditions are identical with their beam counterparts except those involving the shear force. By (3.1.0) we see that

$$
F = M' + Pw' = EI(w''' + \alpha^2 w'), \qquad (3.1.8)
$$

and the $\alpha^2 w'$ term we see is additional to the beam expressions. More realistic conditions in practice might be support moment proportional to slope change and such conditions can be simply formulated and systematically included.

The reader can easily examine (3.1.6) and attach meanings to the constants A–D for himself. For emphasis however we note that B is $-M_{(0)}/P$ and C is $F_{(0)}/P$; namely B and C have immediate physical interpretations in terms of the bending moment ($M_{(0)}$) and shear force ($F_{(0)}$) at the origin. Further $w_{(0)} = B + D$ and $w'_{(0)} = \alpha A + C$. Note particularly though that it is C and not A which is essentially the origin shear force. This is in contrast to the beam where $A(\times 6\beta)$ is the origin shear force. The three constants A, B, D have the dimensions of a displacement, that is a length. The constant C is dimensionless.

3.2 The physical phenomena: buckling and deflexion problems

The equation (3.1.3) and the solution (3.1.6) together describe two physically distinct structural responses to load. In the first case if the equation *and* the boundary conditions are *homogeneous*, that is have zero right-hand sides, physically, if no load or disturbance such as a lack of straightness is applied to the column except at its ends, and then only such as to produce zero boundary conditions, then the theory predicts the column to remain straight and apparently unaffected until an initial critical axial load is reached at which load the column suddenly bends and for most practical purposes loses its structural usefulness. The column is said to buckle; the mathematical problem is called an *eigenvalue* problem. There is usually a spectrum of distinct eigenvalues, the first of the ascending series of eigenvalues gives the buckling load and the associated eigenvector defines the shape into which the column bends. The higher eigenvalues correspond to buckling in higher, more curved, modes, which modes are physically unattainable except when the column is supported so as to prevent the lower modes from developing. The amplitudes of the modes are arbitrary. This is an essential feature of the theory.

In contrast, in the second category of problem, there is some non-homogeneity present, either in the loads – loads other than end load present – or there is an initial lack of straightness giving a particular integral for all values of P. In such cases the column deflects for all values of P and continues to deflect with increasing P but with a final rapid increase in the displacement as the end load approaches the initial eigenvalue load. The problem is then just another stiffness problem such as encountered with beams. There is no arbitrariness in the displacements.

A given strut length with prescribed end conditions possesses an associated eigenvalue problem and this is found by posing the question – for what values of the end load, P, can the initially straight strut assume a *displaced* form. The earliest classical investigations in this field were made by Leonard Euler, a Swiss in the employ of the King of Prussia, who in 1743 studied the problem of the uniform pin ended strut. The associated lowest eigenvalue is known as the Euler load and is to this day the most important single result in the entire field of stability studies.

The Euler theory was for a century or more not seriously studied by engineers. But today, although elaborated in various ways for particular purposes and types of material, it forms the foundation of modern studies.

3.3 The single column length: Euler theory: homogeneous problem

The prototype buckling problem is the Euler column, defined by (3.1.6) with P.I. = 0 and $w = w'' = 0$ at $x = 0, L$. These conditions describe a column of length L, pinned at the two ends and subjected to end load P.

Now $w = A \sin \alpha x + B \cos \alpha x + Cx + D$ and, with an origin at one end, $w = w'' = 0$ at $x = 0$ give

$$\left. \begin{array}{l} B + D = 0, \\ -\alpha^2 B = 0, \end{array} \right\} \tag{3.3.0}$$

from which we conclude that

$$B = D = 0. \tag{3.3.1}$$

The conditions $w = w'' = 0$ at $x = L$ then give

$$\left.\begin{array}{r} A \sin \alpha L + CL = 0, \\ -\alpha^2 A \sin \alpha L = 0. \end{array}\right\} \tag{3.3.2}$$

Now $\alpha \neq 0$, since $P \neq 0$, hence $C = 0$ and

$$\sin \alpha L = 0, \tag{3.3.3}$$

in order that $A \neq 0$, since otherwise the column would remain straight.

The eigenvalue condition (3.3.3) requires that

$$\alpha L = n\pi \quad (n = 1, 2, \ldots). \tag{3.3.4}$$

The case $n = 1$ gives

$$P_{cr} = \frac{\pi^2 EI}{L^2}, \tag{3.3.5}$$

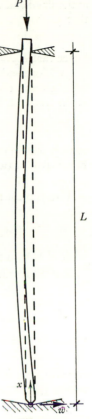

Fig. 3.3.0

which is the *Euler critical load*. In the language of eigenvalue problems, it is the lowest eigenvalue and the associated column eigenvector is $(A, 0, 0, 0)^T$, implying a deflected shape of a sine curve as indicated in fig. 3.3.0, but with arbitrary amplitude A. (The eigenvector is the vector of coefficients (A, B, C, D). The superscript T is used to mean transpose, that is a column rather than a row vector as written.) This arbitrariness is an essential feature of all eigenvalue problems, and is a reflexion of the fact that the theory is a small deflexion theory and large displacements are likely to develop at the critical load.

Exercises

(1) In 1778 Euler published his studies on the fixed ended column length. Show that in the present notation this problem can be described by the boundary conditions $w = w' = 0$ at $x = 0, L$. Show further that the buckling load is given by the lowest value of α from $\cos \alpha L = 1$ and hence that

$$p_{cr} = 4\pi^2 EI/L^2,$$

that is four times the pin-end value. Hence show that the associated eigenvector is given by $D(0, -1, 0, 1)^T$.

(2) If one end is fixed and one pinned show that the buckling load is given by the lowest root of $\tan \alpha L = \alpha L$ with an associated eigenvector for an origin

at the fixed end of $D(1/\alpha L, -1, -1/L, 1)^T$. Evaluate the lowest root and show that

$$\alpha L \doteq \frac{3\pi}{2} - \frac{2}{3\pi}$$

hence that $P_{cr} = 2 \cdot 05(\pi^2 EI/L^2)$.

(3) If one end is free and one fixed, namely a cantilever, show that $\cos \alpha L = 0$, $\alpha L = \frac{1}{2}\pi$ and the associated eigenvector is $A(-1, 0, 0, \sin \alpha L)^T$, for an origin at the free end. Hence

$$P_{cr} = \frac{\pi^2 EI}{4L^2},$$

namely one quarter the Euler pin-end value, since the strut is behaving like an Euler strut of twice the actual length.

The typical result of any of the above calculations is

$$P_{cr} = K\frac{\pi^2 EI}{L^2}, \tag{3.3.6}$$

with $K = 1$ in the Euler case.

It is convenient to rearrange the result for purposes of understanding its significance. Now $EI = (EI)$ minimum, then write

$$I_{\min} = \mathscr{A}k^2, \tag{3.3.7}$$

which is the equation defining the minimum radius of gyration (k), $\mathscr{A}$ being the area of the section. For the useful range of loads on the column, the force P can be thought of as giving rise to a uniform axial stress such that $P = \mathscr{A}\sigma$ and hence we may define σ_{cr} as $P_{cr} = \mathscr{A}\sigma_{cr}$ whence

$$\sigma_{cr} = \frac{K\pi^2 E}{(L/k)^2}. \tag{3.3.8}$$

Plotted on axes of σ_{cr} and (L/k) this result (3.3.8) appears as on fig. (3.3.1). The primary parameter in the study of column instability is therefore the L/k, the so-called *slenderness ratio*.

In practical circumstances there are limits on L/k below which the column crushes without instability and above which the carrying capacity, $\mathscr{A}\sigma_{cr}$, is too small for any practical strut. Hence we derive a simple proto-type strut curve such as ABC (fig. 3.3.1). Other practical effects such as lack of straightness and inelastic effects tend to produce a working curve to which ABC is an upper ideal limit.

Note

3.1 In any given problem we are free to choose an origin of coordinates. For ease of computation if a pin or a free end exists in a problem this should be

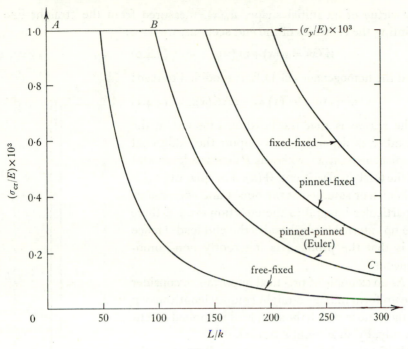

Fig. 3.3.1

chosen as origin. The effect is to put two of the group of four constants A–D immediately to zero. If a pin-end is chosen as origin then $B = D = 0$, if a free end is chosen as origin then $B = C = 0$. Alternatively if an origin on a line of symmetry is chosen then $A = C = 0$, if on a line of antisymmetry $B = D = 0$.

3.4 The single column length: non-homogeneous problems

The typical non-homogeneous problem arises when the strut is imperfect, for example if not entirely straight. Clearly this is likely to be a common practical happening. The effect, as we have remarked in 3.0, is to produce a *deflexion* rather than an eigenvalue problem.

In 3.3 the w displacement is the *total* transverse displacement measured from the line of action of the end load and this is assumed zero for all points $0 \leqslant x \leqslant L$ when $P < P_{cr}$. It is also the *additional* displacement from the initial position.

In the present range of problems we must be careful to distinguish between these two measures. Let $W(x)$ denote the *total* displacement and $w(x)$ the *additional* displacement. Then if the column length is imperfect by virtue of an initial shape, $w_0(x)$, measured from the straight line jointing the centroids of the end sections, we have:

$$W(x) = w(x) + w_0(x), \qquad (3.4.0)$$

and the homogeneous G.D.E. is modified to read:

$$w^{\mathrm{IV}}(x) + \alpha^2 w''(x) = -\alpha^2 w_0''(x). \qquad (3.4.1)$$

The reason is that the bending moment in the member is dependent only upon the additional displacement, $w(x)$, where as P acts at a lever arm of the total displacement, $W(x)$. The quantity $w_0(x)$ is known or assumed at the outset and contributes a particular integral to the equation even if there are no external loads, besides the end load. Hence it is that the problem is inherently non-homogeneous.

As an example of practical importance, consider the uniform initially straight column length shown in fig. 3.4.0 in which the axial load is applied eccentrically, by an amount e, at each end.

With the choice of origin as indicated, the problem is described by

$$\left. \begin{aligned} w &= 0; \quad x = 0, L, \\ w'' &= \frac{Pe}{EI}; \quad x = 0, L. \end{aligned} \right\} \qquad (3.42.)$$

In this case the non-homogeneity enters through the boundary conditions. Now

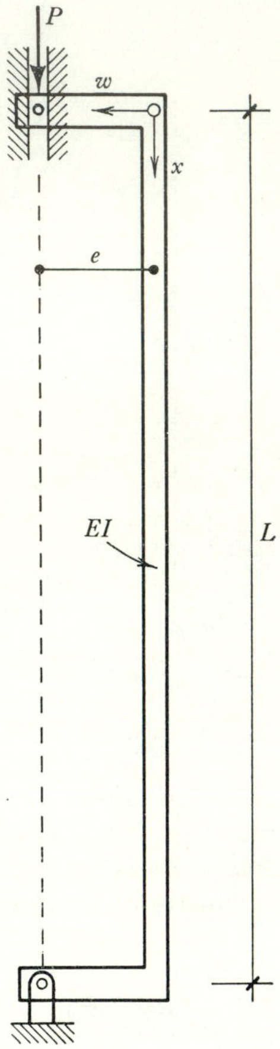

$$w = A \sin \alpha x + B \cos \alpha x + Cx + D,$$

and

$$w = 0, \quad w'' = \alpha^2 e, \quad x = 0$$

give

$$\left. \begin{aligned} B + D &= 0, \\ -\alpha^2 B &= \alpha^2 e, \end{aligned} \right\} \qquad (3.4.3)$$

whence

$$B = -e = -D. \qquad (3.4.4)$$

At $x = L$, $w = 0$, $w'' = \alpha^2 e$ give

$$A \sin \alpha L - e \cos \alpha L + CL + e = 0,$$

$$-\alpha^2(A \sin \alpha L - e \cos \alpha L) = \alpha^2 e, \qquad (3.4.5)$$

Fig. 3.4.0

whence
$$A = \frac{e(\cos \alpha L - 1)}{\sin \alpha L}, \quad C = 0. \tag{3.4.6}$$

The maximum (central) displacement is then given by
$$\left.\begin{aligned} w(\tfrac{1}{2}L) &= A \sin \tfrac{1}{2}\alpha L + B \cos \tfrac{1}{2}\alpha L + D \\ &= -e(\sec \tfrac{1}{2}\alpha L - 1). \end{aligned}\right\} \tag{3.4.7}$$

To interpret the result (3.4.7) first form the total displacement by subtracting (3.4.7) from e and hence
$$W(\tfrac{1}{2}L) = e \sec \tfrac{1}{2}\alpha L, \tag{3.4.8}$$

whence it can be seen that the maximum displacement, measured from the line of action of the thrusts is $e \sec \tfrac{1}{2}\alpha L$. This result is known as the *secant strut formula*.

More intricate problems of the elastic single column length are encountered but they are all of either one or other of the two preceding types. We propose now to discuss the discontinuous solutions appropriate to the G.D.E. (3.1.3) and the solution (3.1.6) and so be in a position to discuss efficiently problems of continuity in column lengths, frames, etc.

3.5 The column: discontinuous solutions

Discontinuous solutions for beams have already been discussed and used. For the column the motive to develop such solutions is the same: namely, to develop particular integrals which contain specific discontinuities corresponding to point loads, applied moments, hinges and other physical changes which are encountered. The method of approach is identical to that used for beams but the solutions are a little more complex by virtue of the $\sin \alpha x$, $\cos \alpha x$ terms in the column complementary function replacing the x^3 and x^2 terms of the beam C.F.

The column equation has already been used in chapter o to introduce the discontinuity method. Here we shall repeat some of that previous analysis in order to compute the D.P.I. appropriate to a point transverse load, W, applied to a distance a from the origin of the x coordinate. In 0.3 we introduced the notation $[\![w(a)''']\!]$ to mean the discontinuity in $w(x)'''$ at $x = a$. The reader should be thoroughly familiar with the contents of 0.3 before proceeding with the present section. Now the present analysis aims to develop a discontinuous particular integral which when added to the continuous complementary function will contain an isolated discontinuity of strength W in the shear force at $x = a$. As can be checked from fig. 3.1.0.the addition of a W force such as in fig. 2.5.0 will cause a positive jump in the value of F at $x = a$. Hence the requirements are summarized as

$$[w] = [w'] = [w''] = 0, \quad [w'''] = W/EI. \qquad (3.5.0)$$

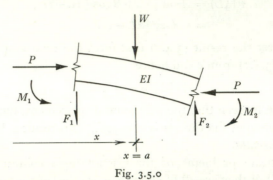

Fig. 3.5.0

Note that although $F = M' + Pw' = EI(w''' + \alpha^2 w')$, since there is to be no discontinuity in slope, w' at $x = a$, then $[w'] = 0$ and hence

$$[w'''] = W/EI.$$

The method now is to exploit the arbitrariness of the conventional continuous P.I. Hence we can write that the

$$\text{D.P.I.} = b \sin \alpha x + c \cos \alpha x + dx + e, \qquad (3.5.1)$$

subject to the four conditions (3.5.0) at $x = a$ for finding the unknown constants b, c, d, e.

Hence

$$\mathbf{D} \begin{bmatrix} b \\ c \\ d \\ e \end{bmatrix} = \begin{bmatrix} 0 \\ 0 \\ 0 \\ W/EI \end{bmatrix}, \qquad (3.5.2)$$

$$\underset{4\times4}{} \underset{4\times1}{} \qquad \underset{4\times1}{}$$

where, from (0.3.6),

$$\mathbf{D} = \begin{bmatrix} \mathscr{S} & \mathscr{C} & a & 1 \\ \alpha\mathscr{C} & -\alpha\mathscr{S} & 1 & 0 \\ -\alpha^2\mathscr{S} & -\alpha^2\mathscr{C} & 0 & 0 \\ -\alpha^3\mathscr{C} & \alpha^3\mathscr{S} & 0 & 0 \end{bmatrix}, \qquad (3.5.3)$$

with $\mathscr{S} = \sin \alpha a, \quad \mathscr{C} = \cos \alpha a.$

By (0.3.7)

$$-\mathbf{D}^{-1} = \frac{1}{\alpha^5} \begin{bmatrix} 0 & 0 & \alpha^3\mathscr{S} & \alpha^2\mathscr{C} \\ 0 & 0 & \alpha^3\mathscr{C} & -\alpha^2\mathscr{S} \\ 0 & -\alpha^5 & 0 & -\alpha^3 \\ -\alpha^5 & \alpha^5 a & -\alpha^3 & \alpha^3 a \end{bmatrix}. \qquad (3.5.4)$$

3.5 *Column: discontinuous solutions*

Now multiply (3.5.4) into the right-hand column vector of (3.5.2) and we have

$$\begin{bmatrix} b \\ c \\ d \\ e \end{bmatrix} = \frac{W}{\alpha^3 EI} \begin{bmatrix} -\cos\alpha a \\ \sin\alpha a \\ \alpha \\ -\alpha a \end{bmatrix}. \tag{3.5.5}$$

The complete particular integral for W is then given by substituting for b, c, d, e from (3.5.5) into (3.5.1) when we obtain:

$$(\text{D.P.I.})_W = \frac{W}{P\alpha}(\alpha z - \sin\alpha z) \quad (z = x - a). \tag{3.5.6}$$

As a check we can see that as $P \to 0$, (3.5.6) yields

$$(\text{D.P.I.})_W \text{ for beam} = \frac{W z^3}{3!\,\beta}, \tag{3.5.7}$$

as is known to be the limiting case for $P = 0$.

Other useful D.P.I.'s can be developed and will be considered shortly. First let us utilize the result (3.5.6) and remember that (3.5.6) is to be set to zero for $z < 0$.

3.6 The single column length: a discontinuous solution

Consider the example illustrated in fig. 3.6.0. We can immediately write down the displaced form, assuming the column initially perfect, as:

$$w = A \sin\alpha x + B \cos\alpha x + Cx + D + (W/P\alpha)(\alpha z - \sin\alpha z) \quad (z = x - a). \tag{3.6.0}$$

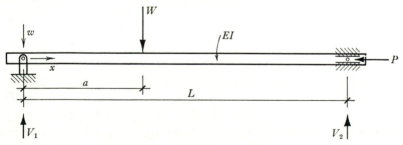

Fig. 3.6.0

The boundary conditions of pin-ends at $x = 0, L$ are then seen to give $B = D = 0$ at $x = 0$ and at $x = L$

$$\left.\begin{array}{l} A \sin \alpha L + CL + (W/P\alpha)(\alpha(L-a) - \sin \alpha(L-a)) = 0, \\ -\alpha^2 A \sin \alpha L + (W/P\alpha)(\alpha^2 \sin \alpha(L-a)) = 0, \end{array}\right\} \quad (3.6.1)$$

whence

$$\left.\begin{array}{l} CL = -(W/P)(L-a), \\ A = (W/P\alpha)(\cos \alpha a - \cot \alpha L \sin \alpha a). \end{array}\right\} \quad (3.6.2)$$

Finally

$$w = \frac{W}{P\alpha}\left((\cos \alpha a - \cot \alpha L \sin \alpha a) \sin \alpha x - \frac{\alpha(L-a)}{L}x + (\alpha z - \sin \alpha z)\right)$$

$$(z = x - a). \quad (3.6.3)$$

Other methods can be used to solve this problem but the present approach appeals as a systematic technique which can be applied in the present and a variety of other circumstances.

3.7   Discontinuous particular integrals: further features

First let us note that it is convenient to think of the D.P.I.s associated with a given G.D.E. as forming a sequence or set. Thus a G.D.E. of order n in terms of a dependent variable w will have associated with it a sequence of D.P.I.s for $[w]$, $[w'] \dots [w^{n-2}]$, $[w^{n-1}] \dots$ In the specific case of the beam, the sequence is better thought of as $[w]$, $[w']$, $[M]$, $[F]$, $[p]$, etc., in the notation of chapter 1.

Now if we seek a D.P.I. for a slope discontinuity in the column case and specify the situation by

$$\left.\begin{array}{l} [w] = [w''] = \dots = [w^{n-1}] = 0, \\ [w'] = \theta, \end{array}\right\} \quad (3.7.0)$$

the result we obtain is   (D.P.I.)$_{w'} = \theta \cdot z$,   (3.7.1)

the same result as for the beam.

However this result is not in the most useful form for application. The reason is that this D.P.I. brings in its train discontinuity in one of the *physical* variables besides the slope, namely the shear force. For, since $F = EI(w''' + \alpha^2 w')$, then $[F] = P[w'] = P\theta$. The expression (3.7.1) ensures that $[w^n] = 0, n \neq 1$ but this is not a sufficient restriction for physical applications. Instead, the requirements for a discontinuous particular integral will be more usefully stated as *continuity* of all the physical variables except the one under consideration, rather than of the sequence $[w^n]$ for all n except a given one. Here w is taken to be the dependent variable in the G.D.E. and in terms of which, therefore, all the physical variables can be expressed as derivatives of w or linear combinations of such derivatives.

In the present case then if for the column we require slope discontinuous by an amount θ but displacement, bending moment and shear force and all higher derivatives continuous, then the necessary D.P.I. must ensure that $[w'] = \theta$ and $[w'''] + \alpha^2[w'] = 0$ with $[w^n] = 0$, $n \neq 1, 3$ at $x = a$, $z = 0$. As can easily be verified, the D.P.I. satisfying these conditions is

$$(\text{D.P.I.})_{w'} = \theta \frac{\sin \alpha z}{\alpha}. \tag{3.7.2}$$

A further feature shared by many members in a given set of D.P.I.s can now be noted. If the parent set of O.D.E.s (and hence the G.D.E.) is of constant coefficient type and is not in *part* integrable in quadratures, then with a usually evident redefinition of the given discontinuity, it can be seen that any one D.P.I. can be found from its neighbours in the sequence by either differentiation or integration.

For example, the D.P.I. for an external (anticlockwise) moment M applied to a column at $x = a$, $z = 0$ is

$$(\text{D.P.I.})_M = (\text{D.P.I.})_2 = (M/P)(1 - \cos \alpha z). \tag{3.7.3}$$

Here the notation $(\ldots)_2$ is used to indicate that the result is one of the sequence $0, 1, 2, \ldots$, dealing basically with discontinuities in the zeroth, first, second, etc., derivatives of the underlying (displacement) variable w. Then if it is imagined that $M = W dz$, M being replaced by equal and opposite forces W acting at the ends of a lever arm dz and then a superposition of Ms from $z = 0$ to z is made; at $z = 0$ an out-of-balance force W will appear but at all subsequent points there will be a cancellation. Hence it is to be expected that

$$(\text{D.P.I.})_W = (\text{D.P.I.})_3 = \int_0^z (W/P)(1 - \cos \alpha z)\, dz$$

$$= \frac{W}{P\alpha}(\alpha z - \sin \alpha z), \tag{3.7.4}$$

as is already known to be the case.

Similarly with $W = p \, dz$,

$$(\text{D.P.I.})_p = (\text{D.P.I.})_4 = \int_0^z \frac{p(\alpha z - \sin \alpha z)}{P\alpha} \, dz$$

$$= \frac{p}{2P\alpha^2} (\alpha^2 z^2 + 2 \cos \alpha z - 2). \qquad (3.7.5)$$

Alternatively, if $M \, dz = EI\theta$, $[\![w']\!] = \theta$ then

$$(\text{D.P.I.})_\theta = (\text{D.P.I.})_1 = \frac{d}{dz} \{ (\text{D.P.I.})_2 \, dz \}$$

$$= \frac{EI}{P} \alpha\theta \sin \alpha z = \frac{\theta \sin \alpha z}{\alpha}, \qquad (3.7.6)$$

as has already been pointed out in (3.7.2).

But it is not true that the $(\text{D.P.I.})_0$, the displacement discontinuity, can be obtained from $(\text{D.P.I.})_1$ in this way. This is because the column G.D.E. *is* in part integrable in quadratures.

Put another way, it will be noted that these properties referred to above are preserved only for discontinuity in physical variables as opposed to mathematical variables. In the present case w''' is not a physical variable although $F = EI(w''' + \alpha^2 w')$ is, and if a mathematical choice of variable is made, say $\quad (\text{D.P.I.})_i = [\![w^i]\!], \quad [\![w^n]\!] = 0, \quad n \neq i,$

then in general the integral and differential properties are destroyed (in part at least).

To summarize, if the (D.P.I.)s are chosen to represent discontinuity in a natural sequence of physical variables and the G.D.E. is of constant coefficient type and *not partially* integrable in quadratures, then these simple differential and integral properties can be seen to hold true. The column equation $w^{\text{iv}} + \alpha^2 w'' = P/EI$ is twice integrable in quadratures and as a result $(\text{D.P.I.})_0$ is not obtainable by differentiation of $(\text{D.P.I.})_1$, as has already been noted. The beam equation $w^{\text{iv}} = p/\beta$ on the other hand is *totally* integrable in quadratures and the entire sequence of (D.P.I.)s can be obtained from any one by differentiation and integration. The equation $w^{\text{iv}} + kw = p/\beta$, the beam on elastic foundation, is not integrable in quadratures at all but the same is true as for beams.

The G.D.E. too must be of constant coefficient type for these properties to hold. Thus the equations for axisymmetric problems of plate bending and stretching possess no such simple relations between the various $(\text{D.P.I.})_i$. The reason is that the $(\text{D.P.I.})_i$ are not expressible in terms of the variable z. For such equations each new (D.P.I.) must be calculated by the methods outlined in chapter o.

3.7 *D.P.I.'s: further features*

There are some further subtleties not referred to explicitly in the present discussion. They relate to the fact that a given G.D.E. cannot be uniquely resolved into a set of lower order equations. Expressed alternatively, a given G.D.E. can represent more than one physical problem. We shall leave the reader to explore these matters as occasion demands.

The column problem possesses one feature which restricts the use of discontinuous solutions as compared with the beam and frame problems already considered. This is the *non-linear* involvement of the axial thrust, P, and section stiffness, EI. Because these quantities occur in the functions comprising the C.F., use of discontinuous solutions in the manner previously demonstrated for frames cannot be extended to columns where there are changes of either P or EI. In such cases choice of an additional origin is forced upon us. But in all other respects the general principles and philosophy of discontinuous solutions applies equally to the column as to the beam and the stable framework. In some circumstances P and EI may change together in such a way as to keep P/EI constant. In these cases full advantage of discontinuous behaviour is obtained with the column as for the beam.

3.8   An example

Determine the lowest critical load for the two span column (fig. 3.8.0). The P and EI are common to both lengths of column and hence a single coordinate origin is sufficient. We make the choice as indicated at A on fig. 3.8.0. Then all possible displaced forms are contained in the expression

$$w = A \sin \alpha x + B \cos \alpha x + Cx + D + (R/P\alpha)(\alpha z - \sin \alpha z), \quad (3.8.0)$$

where $\alpha^2 = P/EI$, R is the reaction at B as shown on fig. 3.8.0, and $z = x - L$.

There are five unknowns in (3.8.0), A, B, C, D, R. The five conditions are two boundary conditions at each end and the zero displacement condition at B. In turn they are

$$w_{(0)} = B + D = 0, \quad w''_{(0)} = -\alpha^2 B = 0; \quad B = D = 0, \quad (3.8.1)$$

$$w_{(L)} = A \sin \alpha L + CL = 0, \quad (3.8.2)$$

$$w_{(3L)} = A \sin 3\alpha L + 3CL + (R/P\alpha)(2\alpha L - \sin 2\alpha L) = 0, \quad (3.8.3)$$

$$w''_{(3L)} = -\alpha^2 A \sin 3\alpha L + (R/P\alpha)(\alpha^2 \sin 2\alpha L) = 0. \quad (3.8.4)$$

Hence $3CL + (2RL/P) = 0$ by subtraction from (3.8.3, 4) and then in (3.8.4)

$$- A \sin 3\alpha L - \frac{3}{2} \left(-\frac{A \sin \alpha L}{L} \right) (\sin 2\alpha L)$$

$$= 0,$$

or

$$2\alpha L (\cot \alpha L + \cot 2\alpha L) = 3. \quad (3.8.5)$$

Now BC is the weaker of the two members AB, BC treated separately, but BC is stronger than a pure Euler strut of the same length. Hence

$$\frac{\pi^2 EI}{L^2} > P > \frac{\pi^2 EI}{4L^2}. \quad (3.8.6)$$

The reader should verify that

$$P \doteqdot 3 \cdot 31 EI/L^2$$

is a good estimate for the first critical load.

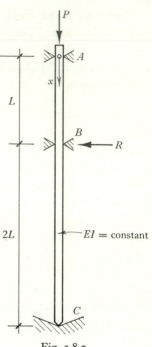

Fig. 3.8.0

3.9 Column stiffness problems: stability functions

The stability problem for the column is the counterpart of the strength problem for the beam. The counterpart of the beam stiffness problem is the solution of the non-homogeneous column problem where the end product is a relation between forces acting, especially the end load, and the displacements produced. If we take a given structure and make two studies, the one with axial load effects included and the other without, then the stiffness of the structure in the former case, the ratio typical force acting/typical displacement produced, is less than in the latter case. That is to say, the primary influence of axial forces in structures at load levels below the lowest critical load is such as to cause a decrease, a deterioration, in stiffness. It is in the load ranges somewhat below the elastic critical load that we are here primarily concerned.

Consider the member AB, initially straight and stress free and finally subject to the force system shown in fig. 3.9.0. The object of our study is to derive relations for the M_{ij} and θ_i. From this information we can derive useful stiffness parameters, the *stability functions* to be defined shortly.

3.9 *Column stiffness problems: stability functions*

Now, as always for a uniform column length:

$$w = A \sin \alpha x + B \cos \alpha x + Cx + D, \qquad (3.9.0)$$

and the boundary conditions here are:

$$w_{(0)} = 0, \quad w''_{(0)} = -M_{12}/EI; \quad w_{(L)} = \Delta, \quad w''_{(L)} = M_{21}/EI. \quad (3.9.1)$$

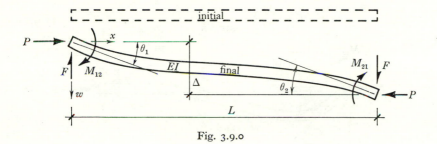

Fig. 3.9.0

Applying (3.9.1) to (3.9.0) we have

$$B + D = 0, \quad -\alpha^2 B = -M_{21}/EI,$$

whence
$$B = -D = M_{12}/P. \qquad (3.9.2)$$

And
$$\Delta = -(M_{12}/P)(1 - \cos \alpha L) + A \sin \alpha L + CL, \qquad (3.9.3)$$

$$\frac{M_{21}}{EI} = -\alpha^2 \left(\frac{M_{12}}{P} \cos \alpha L + A \sin \alpha L \right), \qquad (3.9.4)$$

whence
$$A = -(M_{21} + M_{12} \cos \alpha L)/P \sin \alpha L, \qquad (3.9.5)$$

$$C = \frac{\Delta}{L} + \frac{M_{12} + M_{21}}{PL}. \qquad (3.9.6)$$

Finally, we have

$$\theta_1 = w'_{(0)} = \alpha A + C$$

$$= \frac{\Delta}{L} - \frac{M_{21}(\alpha L - \sin \alpha L)}{PL \sin \alpha L} + \frac{M_{12}(\sin \alpha L - \alpha L \cos \alpha L)}{PL \sin \alpha L} \qquad (3.9.7)$$

and
$$\theta_2 = w'_{(L)} = \alpha A \cos \alpha L - \alpha B \sin \alpha L + C$$

$$= \frac{\Delta}{L} - \frac{M_{12}(\alpha L - \sin \alpha L)}{PL \sin \alpha L} + \frac{M_{21}(\sin \alpha L - \alpha L \cos \alpha L)}{PL \sin \alpha L}. \qquad (3.9.8)$$

Note the symmetry for exchange of suffices 1, 2 in (3.9.7, 8).

We now wish to specialize the deformation depicted in fig. 3.9.0 to the case $\Delta = \theta_2 = 0$ (fig. 3.9.1), and then define S, C the stiffness and carry-over factors respectively by the relations

$$M_{12} \equiv Sk\theta_1 \quad (k = EI/L), \tag{3.9.9}$$

$$M_{21} \equiv CM_{12}. \tag{3.9.10}$$

Then from (3.9.8)

$$C = \frac{M_{21}}{M_{12}} = \frac{\alpha L - \sin \alpha L}{\sin \alpha L - \alpha L \cos \alpha L}. \tag{3.9.11}$$

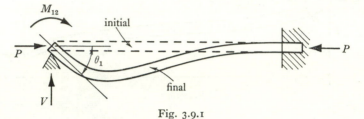

Fig. 3.9.1

Now using (3.9.9, 10, 11) in (3.9.7) there follows that

$$\theta_1 = \frac{Sk\theta_1}{PL \sin \alpha L} \left(\frac{-(\alpha L - \sin \alpha L)^2 + (\sin \alpha L - \alpha L \cos \alpha L)^2}{\sin \alpha L - \alpha L \cos \alpha L} \right),$$

or

$$PL \sin \alpha L = Sk(1 - C^2)(\sin \alpha L - \alpha L \cos \alpha L), \tag{3.9.12}$$

a very useful relation. Now use (3.9.11) in (3.9.12), when

$$S = \frac{\alpha L(\sin \alpha L - \alpha L \cos \alpha L)}{2(1 - \cos \alpha L) - \alpha L \sin \alpha L}. \tag{3.9.13}$$

The more usual notation is to put $\alpha L = 2\beta$ when

$$\left. \begin{array}{l} S = \dfrac{\beta(1 - 2\beta \cot 2\beta)}{\tan \beta - \beta}, \\[3mm] C = \dfrac{2\beta - \sin 2\beta}{\sin 2\beta - 2\beta \cos 2\beta}. \end{array} \right\} \tag{3.9.14}$$

A second situation met with in practice is the case when $\Delta = M_{21} = 0$ (fig. 3.9.2). In this case we define another stiffness S^* such that

$$M_{12} \equiv S^* k\theta_1, \tag{3.9.15}$$

and then, from (3.9.7)

$$\theta_1 = \frac{M_{12}(\sin \alpha L - \alpha L \cos \alpha L)}{PL \sin \alpha L}, \tag{3.9.16}$$

and on using (3.9.12) we obtain

$$M_{12} = Sk(1 - C^2)\,\theta_1 \quad \text{or} \quad S^* = S(1 - C^2). \qquad (3.9.17)$$

These functions S, C, S^* are tabulated in tables 3.9.0, 1 (appendix) for a range of values of $p = P/P_E$; $P_E \ (= \pi^2 EI/L^2)$ being the Euler load for the given column length. This is a convenient parameter and is related to α and β of (3.9.13, 14) by:

$$\alpha^2 = P/EI = (\pi/L)^2 p, \qquad (3.9.18)$$

$$\beta^2 = (\tfrac{1}{2}\pi)^2 p. \qquad (3.9.19)$$

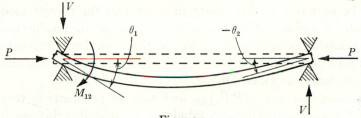

Fig. 3.9.2

Exercises

(1) Analyse the M/Δ relation for the situation shown in fig. 3.9.3. Use any results already established, and especially (3.9.12), and hence show that

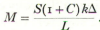

$$M = \frac{S(1 + C)\,k\Delta}{L}.$$

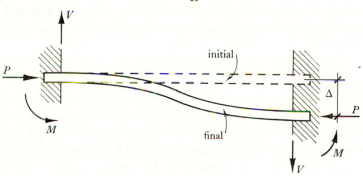

Fig. 3.9.3

(2) Specialize the expressions for S, C, S^* to the case $P \to 0$. Hence show that

$$S(0) = 4, \quad C(0) = \tfrac{1}{2}, \quad S^*(0) = 3.$$

(3) Show that (3.9.7, 8) can be written in terms S, C as

$$\theta_1 = \frac{M_{12} - CM_{21}}{Sk(1 - C^2)}, \quad \theta_2 = \frac{M_{21} - CM_{12}}{Sk(1 - C^2)}.$$

131

3.10   Column stiffness: some applications of stability functions

It may seem a somewhat curious situation to note that the main applications for the 'stability' functions defined in 3.9 is in tackling stability rather than stiffness problems. The point to appreciate is however that a definition for instability is that the structure becomes *unstable* when it offers vanishingly small *stiffness* to applied load or disturbing influence. For ease of calculation the simplest applied load or disturbing influence is chosen – typically an applied external joint moment or force. Some care must be exercised in this choice in order that the correct mode of instability is initiated. With this safeguard observed, the results obtained from a stability function analysis are identical with those obtained from a first principles eigenvalue analysis. To show this let us re-examine the problem already considered in 3.8.

The values for $p_{AB} = (P/P_E)_{AB}$, and p_{BC}, the parameter in terms of which the stability functions are tabulated, are related in this case by

$$p_{BC} = 4p_{AB}. \qquad (3.10.0)$$

Now consider a test moment applied at B. This moment will divide between BA and BC according to the respective S^* values. We do not need to know this information however. We need merely observe that as the critical P value is reached, the structure will offer less and less resistance to the applied moment until at the critical load the stiffness to M will be zero. This condition is expressed by the relation

$$S^*_{BA} + S^*_{BC} = 0. \qquad (3.10.1)$$

Hence the value of P can be found if we use table 3.9.0 (appendix) to find a p_{BC} which will satisfy (3.10.0) and (3.10.1). It is clear that the S^*'s must be of opposite sign and hence the larger of the p's, namely p_{BC}, must be greater than unity. Glancing down the table $p_{AB} = 0.3$, $p_{BC} = 1.2$ will give a left hand side in (3.10.1) to be approx. $2.35 - 1.17 = 1.18$ whereas $p_{AB} = 0.4$, $p_{BC} = 1.60$ will give $2.10 - 6.03 = -3.93$. Hence the required value lies within these two limits. Proceeding thus it is a simple matter to find $p_{AB} \doteqdot 0.335$ to be a satisfactory solution giving

$$P = \frac{0.335\pi^2 EI}{L^2} = \frac{3.31 EI}{L^2}. \qquad (3.10.2)$$

As a second example consider the continuous column shown in fig. 3.10.0. Our aim is to use the stability functions tabulated in table 3.9.0

to find P_{cr} for this column. The process requires us to imagine some deformed shape – in this case we suppose the joints B and C to rotate as indicated, carrying the column into the shape indicated. We have chosen θ_B to be negative because our convention for rotations is positive clockwise.

The following relationships can now be written down from the definitions for S and C introduced in 3.9 and Exercise 3.

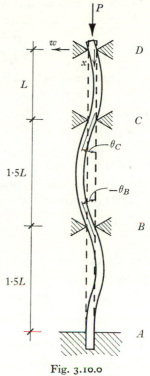

$$M_{CD} = S^*_{CD} k_{CD} \theta_C, \qquad (3.10.3)$$

$$\theta_C = \frac{M_{CB} - C_{CB} M_{BC}}{S_{CB} k_{BC} (1 - C^2_{CB})}, \qquad (3.10.4)$$

$$\theta_B = \frac{M_{BC} - C_{BC} M_{CB}}{S_{BC} k_{CB} (1 - C^2_{BC})}, \qquad (3.10.5)$$

$$M_{BA} = S_{BA} k_{BA} \theta_B. \qquad (3.10.6)$$

Now $S_{BC} = S_{CB} = S_{BA}$ and $C_{CB} = C_{BC}.$ Also

$$\left. \begin{array}{l} M_C = M_{CD} + M_{CB} \to 0, \\ M_B = M_{BC} + M_{BA} \to 0, \end{array} \right\} \quad (3.10.7)$$

Fig. 3.10.0

at instability.

Then (3.10.3–7) are six homogeneous equations in six unknowns θ_B, θ_C, M_{BA}, M_{BC}, M_{CB}, M_{CD}. If the solutions are not all to be just zero then the determinant of the coefficient matrix of this set of equations must be zero. Alternatively, quite simply, if we eliminate M_{CD} through (3.10.3) and the first of (3.10.7), then eliminate θ_C from the resulting equation and (3.10.4) we obtain

$$S^*_{CD} k_{CD} (M_{CB} - C_{CB} M_{BC}) = -S_{CB} k_{BC} (1 - C^2_{CB}) M_{CB}. \qquad (3.10.8)$$

In a similar fashion for M_{BA} and θ_B we obtain

$$S_{BA} k_{BA} (M_{BC} - C_{BC} M_{CB}) = -S_{BC} k_{BC} (1 - C^2_{CB}) M_{BC}. \qquad (3.10.9)$$

Now $k_{BA} = EI/1{\cdot}5L = k_{BC}$. Hence, finally, by forming up the ratio M_{BC}/M_{CB} from each of (3.10.8 and 9) and equating we obtain

$$1 + \frac{S_{BA}(1 - C^2_{BC})}{1{\cdot}5 S^*_{CD}} = \frac{C^2_{BC}}{2 - C^2_{BC}}. \qquad (3.10.10)$$

Here we have used that $k_{CD} = 1\cdot5k_{BC}$. Now P is common to all three column lengths and hence:

$$p_{CD} = \frac{P}{\pi^2 EI/L^2} = \frac{P}{(P_E)_{CD}} = \frac{1}{2\cdot25}\frac{P}{(P_E)_{BC}} = \frac{p_{BC}}{2\cdot25}. \qquad (3.10.11)$$

We are hence in a position to use table 3.9.0 to find a value for p_{BC}.

If $p_{CD} = 1/1\cdot5$, $p_{BC} = 1\cdot5$, then $S_{BC} \doteq 1\cdot46$, $C_{BC} \doteq 1\cdot97$, $S^*_{CD} \doteq 1\cdot32$. Hence L.H.S.–R.H.S. (3.10.10) $= -1\cdot14+2\cdot06 = 0\cdot72$.

If $p_{CD} = 0\cdot71$, $p_{BC} = 1\cdot6$, then $S_{BC} \doteq 1\cdot22$, $C_{BC} \doteq 2\cdot43$, $S^*_{CD} \doteq 1\cdot19$. Hence L.H.S.–R.H.S. (3.10.10) $= -2\cdot38+1\cdot51 = -0\cdot87$.

A linear interpolation gives $p_{BC} = 1\cdot55$ and this is a sufficiently accurate value for practical purposes.

The critical load is then given by

$$P = \frac{1\cdot55\pi^2 EI}{(1\cdot5)^2 L^2} = 6\cdot78\frac{EI}{L^2}, \qquad (3.10.12)$$

namely the strength of the complete column is approximately 50 % greater than the strength of the weakest segment, BC, treated as an Euler column.

3.11 The tie: increased stability

The whole of the foregoing part of this chapter is concerned with P, the axial force, compressive. If P is tensile, the eigenvalue problem for the perfect member disappears – there are no instability phenomena – and for the imperfect member the presence of P enhances the stability and stiffness.

Now the G.D.E. is $\qquad w^{iv} - \alpha^2 w'' = 0, \qquad (3.11.0)$
and the solution is

$$w = A\sinh\alpha x + B\cosh\alpha x + Cx + D. \qquad (3.11.1)$$

All of the discussion of column stiffness problems can be translated into tie stiffness problems and all the discussion of discontinuous solutions can likewise be modified. In table 3.9.1 are tabulated values for S, C, S^* in the case of P a tension. We shall not pursue the topic further here.

3.12 Inelastic column theory

In all our discussion thus far a proportionality between stress and strain (or better moment and curvature) has been assumed. Namely, the material has been assumed to be elastic. We are unable however to give a convincing simple theory of plastic or elastic–plastic column behaviour in which the moment/curvature relation shows the level plateau typical of the M/κ relations of fig. 1.0.0. But there is a further type of stress/strain behaviour, typical of alloy materials, which can be adequately dealt with and is important in practical applications. Consider the stress/strain relation fig. 3.12.0 which is typical of aluminium alloy materials tested

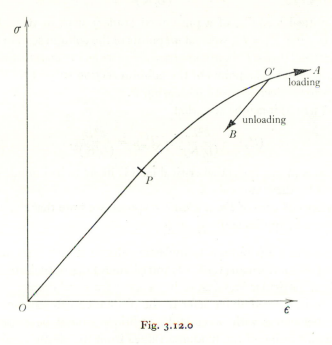

Fig. 3.12.0

in simple tension. The salient features of the response are a gradually decreasing modulus, measured as $\Delta\sigma/\Delta\epsilon$, as σ increases, the lack of a yield point and an inherent irreversibility. Thus $O'A$ is a typical segment of σ/ϵ for *loading* and $O'B$ for *unloading*, $O'B$ being parallel to OP, P being the limit of proportionality. We speak of this as 'elastic unloading'.

Associated with a typical point O' are two moduli, the tangent and reduced modulus respectively. Thus the tangent modulus E_t is given by

$$E_t \equiv d\sigma/d\epsilon \quad (\sigma \text{ increasing}), \tag{3.12.0}$$

namely a measure of the curve slope at O'. The reduced modulus E_r is section shape dependent and has the value

$$E_r \equiv \frac{4E_t E}{(\sqrt{E}+\sqrt{E_t})^2}, \tag{3.12.1}$$

for the rectangular section. The modulus E is the slope of portion OP of the σ/ϵ curve. We note that

$$E_t \leqslant E_r \leqslant E. \tag{3.12.2}$$

These moduli are used in expressions of the form

$$P_{cr} = \frac{\pi^2 E_i}{(L/K)^2} \tag{3.12.3}$$

for the critical load P_{cr}, of a pin-ended (Euler) strut in the following situations. First, $E_i = E_t$, when at *all* points of the column section subject to moment and thrust P there is *no* stress decrease as P increases. Secondly, $E_i = E_r$, when at *some* points of the column section subject to M and P there *is* a stress decrease with increasing P.

From (3.12.2) it can be seen that

$$(P_{cr})_r = \frac{\pi^2 E_r}{(L/K)^2} > (P_{cr})_t = \frac{\pi^2 E_t}{(L/K)^2}. \tag{3.12.4}$$

For real columns the actual critical load is likely to be bounded above and below by these two values.

In the special case of *linear elastic* response we note that the equality sign in (3.12.2) applies and $E_t = E_r = E. \tag{3.12.5}$

The pioneer contributions to inelastic column studies were made by Engesser (1889), Kármán (1908, 10) and Shanley (1947). Inelastic column theory is of particular importance in aircraft structural design.

The results quoted above refer to the pin-ended column. Results for inelastic behaviour with other end conditions cannot be obtained by simple modification of the modulus values from the elastic results. The specialist literature should be referred to for further information.

3.13 Conclusions

Our primary interest in the present discussion of column behaviour has not been to present an exhaustive treatment but rather to treat a class of problems only, elastic problems, and within this class emphasize the notions of discontinuous solutions.

The student interested in a more extensive treatment of the elastic problems treated in this chapter is particularly recommended to consult Gregory (3.16), whereas the student wishing to study inelastic behaviour should go to Drucker (3.16), chapter 16, the early chapters of Bleich or Galambos (3.16).

3.14 Summary of discontinuous particular integrals for the column

The transverse displacement, slope, bending moment, transverse shear force and uniform distributed load will be denoted, respectively, by w, w', M, F, p. Then, with $\alpha^2 = P/EI$ and $[\ldots]$ to mean 'discontinuity in ...'

$$\left. \begin{array}{l} [w] = \Delta \\ [w'] = \theta \\ [M] = M \\ [F] = W \\ [p] = p \end{array} \right\} \text{implies a D.P.I.} \left\{ \begin{array}{l} = \Delta \cdot \cos \alpha z \\ = \dfrac{\theta \sin \alpha z}{\alpha}, \\ = \dfrac{M}{P}(1 - \cos \alpha z), \\ = \dfrac{W}{P\alpha}(\alpha z - \sin \alpha z), \\ = \dfrac{p}{2P\alpha^2}(\alpha^2 z^2 - 2 + 2\cos \alpha z). \end{array} \right.$$

The D.P.I. when added to the usual C.F. and any P.I.s which are continuous through the value $x = a$, $z = 0$, will give the required general solution for the column length, provided that P and EI do not change discontinuously across the section $x = a$.

3.15 Exercises

The following two problems are left for the reader to attempt. Further examples can be found in the texts listed in the references, 3.16.

(1) For the frame in fig. 3.15.0 show that the frame stiffness to a horizontal force is given by

$$\frac{F}{\Delta} = \frac{W}{L\left(\dfrac{2}{\alpha L}\tan\dfrac{\alpha L}{2} - 1\right)} \quad \text{with} \quad \alpha^2 = \frac{W}{2EI}.$$

Hence deduce that the frame first becomes unstable due to W, in a sidesway mode, when

$$\frac{\alpha L}{2} = \frac{\pi}{2} \quad \text{or} \quad W = \frac{2\pi^2 EI}{L^2}.$$

(2) A uniform elastic column for testing with pin-ends is fitted with rigid end shoes as shown in fig. 3.15.1. The composite member is of overall length L and the end fittings are each of length a.

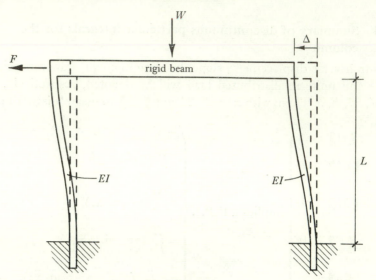

Fig. 3.15.0

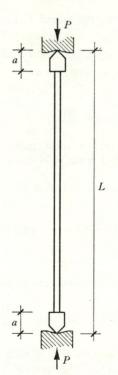

Fig. 3.15.1

Make a study of this column from the classical Euler point of view and show that

$$P_{cr} = \frac{\pi^2 EI}{L^2}\left(1 + \frac{\pi^2\alpha^3}{6}\right), \quad \alpha = \frac{2a}{l}.$$

(3) *The Southwell Plot.* No real column is perfectly straight, and if tested pin-ended for a range of values of end load, P, below the Euler load, the column will deflect progressively and a P/deflexion plot will be obtained. Provided that the column initial lack of straightness is something approaching a simple bowing from end to end, then the following procedure, due to Southwell (3.16), provides a very satisfactory method for displaying the experimental results.

Let us suppose the lack of straightness to be adequately described by an initial lateral displacement from the line joining the end points of

$$w_0(x) = \Sigma \sin(\pi x/L),$$

where the origin for x is an end and the column is pin-ended, of length L.

Hence show that for an axial load $P < P_E$, the *further* lateral displacement $w(x)$, which is the displacement observed in the test, is given by

$$w = \frac{P\Sigma}{P_E - P}, \quad P_E = \frac{\pi^2 EI}{L^2} = \text{Euler load}.$$

The Southwell recommendation is to graph $w/P/w$ when a straight line plot results, with inverse slope P_E and intercept on the w axis of Σ. Hence, experimentally, both P_E and Σ can be simply found and, more importantly, P_E is found as a slope and not an intercept.

3.16 References

The literature on column problems is very extensive and considerable selection is necessary. The following works are suggested as supplementary reading.

Baker, Sir John, Horne, M. R. and Heyman, J. *The Steel Skeleton*, vol. II, Cambridge University Press, London (1956). The 'Columns' chapter is interesting if rather demanding reading. The book as a whole reports at first hand the basis for the rigid–plastic approach to framework design.

Bleich, F. *Buckling Strength of Metal Structures*, McGraw-Hill, New York (1952). This is a balanced summary of the whole range of instability problems up to the 1950s and can be read with interest.

Drucker, D. C. *Introduction to Mechanics of Deformable Solids*, McGraw-Hill, New York (1967). A modern American text. Useful discussion of instability phenomena of column-like members. Problems without answers.

Galambos, T. V. *Structural Members and Frames*, Prentice-Hall, New Jersey (1968). This recent text has a design oriented approach and gives the most complete treatment of the practical column problem of all the references included here. Both elastic and inelastic instability discussed.

Gregory, M. S. *Elastic Stability*, Spon, London (1967). A recent text with a large number of worked examples.

Southwell, R. V. *Theory of Elasticity*, Oxford University Press, London (1936 and 1941). Southwell contributed significantly to the theory in the formative 1910–30 period. His main legacy is the 'Southwell Plot'. The relevant chapters of his book are useful reading.

Shanley, F. R. *Mechanics of Materials*, McGraw-Hill, New York (1967), or *Strength of Materials*, McGraw-Hill, New York (1957). The former is a revision of the latter. Shanley pioneered modern inelastic column theory and both books are full of interesting detail on this subject. Problems with answers (in F.P.S. units) included.

Timoshenko, S. P. and Gere, J. *Theory of Elastic Stability*, McGraw-Hill, New York (1961). This is the revision of Timoshenko's 1936 book of the same title. It has probably had more influence than any other single text in English in popularizing stability problem study. There is only limited discussion of the practical column problem since the subject matter ranges over many other topics besides.

In all the foregoing books there is no adequate mention of discontinuity methods as we have been discussing them. Our final two references draw attention to the discontinuity methods. The second of the two was the starting point for the present book.

Webb, H. A. and Ashwell, D. G. *Mathematical Toolkit for Engineers*, Longmans Green, London (1959), 2nd edition. The first edition of this slim volume did not contain a discussion of the discontinuity method. In the second edition the column with a transverse point load is discussed – although from the standpoint of a second order equation for the bending moment. Also discussed is the beam on an elastic foundation and a continuous column. Herbert Webb (1882–1961) was one of the last of the coaches in the nineteenth century Cambridge Mathematical tradition. He published little but taught a great deal; in particular he taught mathematics to engineers in Cambridge in the 1920's and 30's. Both he and W. H. Macaulay (1853–1936) trained as mathematicians through the Cambridge Tripos. Both, too, spent most of their adult years in Cambridge. Macaulay published his beams paper in 1919 (see 1.20). Webb took up the idea and extended it but published nothing of it until the appearance of the *Toolkit* with Dr Ashwell. W. H. Wittrick, in the next reference, pays tribute to Webb's teaching on this subject.

Wittrick, W. H. 'A generalisation of Macaulay's method with applications in structural mechanics'. *A.I.A.A. Journal* (1965), **3**, 326. In this paper, as has been noted above, Wittrick pays tribute to the teaching and priority of H. A. Webb in putting forward the discontinuity method as applied to boundary value problems. Professor Wittrick discusses the equations $ßw^{iv} + Pw'' + kw = p$ and $D(1/r(rw_r)_r)_r = F$ and provides a very useful summary of the known results and state of the art, adding a number of results of his own as well. There is however no explicit discussion of displacement discontinuities, of multiple discontinuities, of the range of physical occurrence of discontinuities and a number of other topics. This paper was the starting point and stimulus for the methods discussed in the present book.

Wittrick, too, notes the priority of Clebsch, (1.20), in using the methods in a statical context. The page reference he quotes should however be p. 424 not p. 462. The actual reference to use of the discontinuity idea is probably p. 390, Clebsch's Art. 87, where the equations for the beam on n supports are formulated.

Throughout the present book we have used the symbol z to indicate the variable associated with a discontinuity. This same notation was used by Wittrick and was originally proposed by H. A. Webb. From correspondence which Dr Ashwell has made available, it seems clear that the paper which W. H. Macaulay published in 1919, (1.20), appeared several years after the method, as applied to the simple beam problem, had been suggested by Macaulay and taught in the Engineering Laboratory in Cambridge by C. E. Inglis. H. A. Webb was working at the Royal Aircraft Establishment, Farnborough, during the 1914–18 War and relates in the correspondence how he taught the method 'to fellow workers..., all lapped it up and wondered why they hadn't learnt it as undergraduates...'. Webb's later extensions of the method to other situations, situations in which the governing equation is not integrable in quadratures, took place some years later it seems and have been made accessible only through the *Toolkit* and the present paper by W. H. Wittrick.

It is perhaps significant that the majority of American references to W. H. Macaulay misspell his name as Macauley and not a few note his initial as R. instead of W. H. This latter mistake would seem to stem from a similar error in John Case's *Strength of Materials*, Arnold (1925) and the footnote on his p. 225. Finally it might be noted that Wittrick does not include the *Toolkit* in his list of references. I had been unaware of the addition of this material on discontinuities to the second edition of the *Toolkit* until shortly before collecting these references. No doubt the addition of this material had also escaped Professor Wittrick's notice.

This is perhaps the most appropriate place to make mention of a number of related contributions by H. Tottenham, some of which remain unpublished. Tottenham has applied matrix methods in the spirit of the Discontinuity approach to a variety of plate and shell problems. He has coined the term *Matrix Progression* to describe the method but as yet no reference which does justice to the work can be quoted. It is to be hoped that this circumstance will soon be righted.

4

FURTHER TOPICS IN
ONE DIMENSIONAL STRAIGHT
MEMBERED STRUCTURES

4.0 Introduction

Thus far the three basic topics of the beam, the framework and the column, as types of rigid jointed frameworks in bending, have been discussed. The approach has been consistently through a discussion of the differential equations and discontinuity principles. In this chapter are gathered a number of topics, all one-dimensional in the sense of being described by ordinary differential equations, and dealing with a variety of further situations met with in practice.

First discussed are out-of-plane bending problems in which a plane structure is loaded normal to the plane of the structure – the *grillage* problem. Typically this situation is met with in floor support and bridge decking systems. The members of such a framework are generally *twisted* as well as *bent* and this leads naturally to consideration of one dimensional frameworks which are not plane. A simple space framework is considered in illustration.

All the problems thus far considered are homogeneous in the sense that the internal forces are assumed to be zero in the absence of external load. In illustration of *non-homogeneous* problems temperature stresses and intentional prestress, such as used in concrete structures, are considered.

Sway effects, that is the effects of joint displacements, in frameworks have been encountered in chapter 2. Although attention has not been directed to the quantitative effects resulting from sway it is easy to appreciate that sway when it occurs is frequently the major internal stress raising influence. In practical frameworks sway should be avoided or minimized if at all possible. One method to achieve a satisfactory control of sway effects is to cross-brace the basically rectangular rigid frame with light diagonal tension ties. A simple application of such bracing is discussed in 4.6.

The chapter continues with a discussion of composite beam action

4.0 *Introduction*

achieved through some sort of connexion and usually between a steel and a concrete member. The concluding section discusses inextensional cable problems with small dip. In a sense these problems are a degenerate form of tie behaviour as discussed in chapter 3 in which the flexural stiffness is negligible. There are however some differences in the presentation. All the topics are treated at an elementary level but by a similar approach, through differential equations and discontinuity principles, to the earlier chapters.

4.1 Out-of-plane loading: grid frameworks equilibrium and material property equations

Thus far in our discussion we have considered plane frameworks loaded so as to produce displacements in the plane of the framework. This is the practical situation for vertical plane frameworks. We now wish to consider a plane rigid jointed framework under the action of loads *normal* to the plane of the framework. This is the practical situation with a grillage of beams such as is used in bridge decking and floor support systems where there is not a column support under each beam–beam junction.

The beams in such a system are commonly twisted as well as bent and we shall use the two displacement variables w, the transverse displacement measured normal to the beam and now also normal to the plane of the beams, together with θ, the twist about the line of centroids, to characterize the deformation.

The bending and twisting effects for a single beam length are independent of each other, although, clearly, in a system of beams the displacements of one beam will be coupled to the twists of another.

For the single beam the bending effects are described by the equilibrium equations

$$M' = F, \quad F' = p, \tag{4.1.0}$$

and the *elastic* material property equation is the familiar

$$\beta w'' = M. \tag{4.1.1}$$

Hence $\qquad \beta w^{\text{iv}} = p, \quad \text{for} \quad \beta = EI = \text{const.} \tag{4.1.2}$

as has already been met with in chapters 1 and 2.

The twisting effects are described by (fig. 4.1.0) the equilibrium equation

$$T' = 0, \tag{4.1.3}$$

and the elastic material property equation

$$\mathscr{G}\theta' = T. \tag{4.1.4}$$

Hence, if $\mathscr{G}$ = torsional rigidity = μJ = constant, then

$$\mathscr{G}\theta'' = 0, \tag{4.1.5}$$

or
$$\theta = \alpha x + \beta + \text{P.I.} \tag{4.1.6}$$

Here μ is the shear modulus, J the second polar moment of area, $\alpha\mathscr{G}$ the origin torque and β the origin twist.

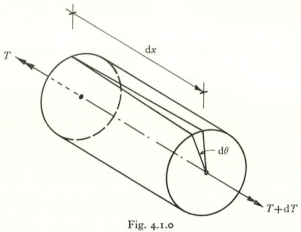

Fig. 4.1.0

In exactly the same manner as for the beam bending equation (4.1.2), so too for the beam twisting equation (4.1.5) and the solution (4.1.6), discontinuous particular integrals can be investigated. In the twisting case in the elastic range, rotation discontinuities are encountered along with a discontinuous torque, typically at a change of direction of a member. The appropriate D.P.I. for a torque $[\![T]\!] = T$ is

$$\frac{Tz}{\mathscr{G}}, \tag{4.1.7}$$

and for a rotation discontinuity $[\![\theta]\!] = \gamma$ the D.P.I. is

$$\gamma(z)^{\circ}. \tag{4.1.8}$$

These and other D.P.I.s can be simply derived by the methods of section 0.3.

We note however that (4.1.5) is a *second* order G.D.E. and of boundary value type and hence requires *one* boundary condition at each end of the domain, not the *two* of the bending equation (4.1.2), which is of *fourth* order.

4.2   Example of elastic bending and twisting

Consider the simple plane rigid jointed framework shown in perspective in fig. 4.2.0, the plane of the members being horizontal and the load vertical. It will be assumed that the beam section has two axes of symmetry arranged in the horizontal and vertical planes. This is the usual case in practice and ensures that vertical loads produce vertical displacements. Let the bending stiffness be denoted by $\beta(=EI)$ and the torsional stiffness by $\mathscr{G}(=\mu J)$, where μ is the shear modulus and J is the polar second moment of area.

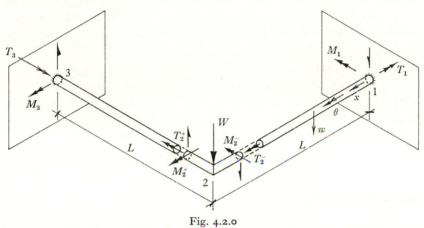

Fig. 4.2.0

We can write down the following displaced forms for transverse displacement and twist (rotation) θ. The origin will be chosen at 1 for both w and θ when

$$w = Ax^3 + Bx^2 + \phi z_2 + \frac{Mz_2^2}{2!\,\beta} + \frac{Wz_2^3}{3!\,\beta}, \qquad (4.2.0)$$

$$\theta = \alpha x + \gamma(z_2)^\circ + \frac{Tz_2}{\mathscr{G}}. \qquad (4.2.1)$$

These descriptions incorporate the conditions at the origin of

$$w = w' = \theta = 0$$

at $x = 0$, by virtue of the absence of terms in Cx and D in (4.2.0) and of β in (4.2.1). Further (4.2.0), (4.2.1) anticipate that the moment and slope for w, and the twist and torque for θ will be discontinuous at 2. The first two of these are analogous to, but in contrast with, the situation at a rigid joint in a plane frame loaded *in* the plane. For in chapter 2 we saw that generally the displacement and shear force are discontinuous at a rigid joint. Here we can see on reflexion about the role of the joint in the present framework that it is the *other* two physical variables which can be expected to be discontinuous, namely slope and bending moment. The discontinuities in θ at a change of direction of the member have already been remarked upon and are easily reasoned.

The situation we are about to discuss however is special in so far that the members at the joint meet at right angles. Then typically twist in one member is transferred to the adjacent member as slope and vice versa, but with care as to signs. Likewise bending moment and torque exchange across the joint. The conditions on M and T can best be obtained by examining joint equilibrium. The free body for the joint is shown on fig. 4.2.0. Finally the shear balance across the loaded joint is already satisfied by use of the previously extensively used expression for

$$[F] = W \text{ as } \frac{W z_2^3}{3!\text{ß}}.$$

To proceed now with the analysis of the framework, we note that there are seven unknowns A, B, ϕ, M; α, γ, T and there are seven conditions to be satisfied.

Geometrical conditions at 2:

$$\left.\begin{array}{l} (w')_{2-} = -(\theta)_{2+}, \\ (\theta)_{2-} = (w')_{2+}. \end{array}\right\} \tag{4.2.2}$$

Equilibrium conditions at 2:

$$\left.\begin{array}{l} M_{2-} + T_{2+} = 0, \\ T_{2-} - M_{2+} = 0. \end{array}\right\} \tag{4.2.3}$$

The usual boundary conditions at 3 of:

$$w = w' = \theta = 0. \tag{4.2.4}$$

We shall now write down the corresponding equations in the same order as enumerated in equations (4.2.2, 3, 4).

Thus

$$\left.\begin{array}{l} 3AL^2 + 2BL = -(\alpha L + \gamma), \\ 3AL^2 + 2BL + \phi = \alpha L, \\ 6A\text{ß}L + 2B\text{ß} + \alpha\mathscr{G} + T = 0, \\ \alpha\mathscr{G} - (6A\text{ß}L + 2B\text{ß} + M) = 0, \\ 8AL^3 + 4BL^2 + \phi L + \dfrac{ML^2}{2\text{ß}} + \dfrac{WL^3}{6\text{ß}} = 0, \\ 12AL^2 + 4BL + \phi + \dfrac{ML}{\text{ß}} + \dfrac{WL^2}{2\text{ß}} = 0, \\ 2\alpha L + \gamma + (TL/\mathscr{G}) = 0. \end{array}\right\} \tag{4.2.5}$$

4.2 Example of elastic bending and twisting

By summing and differencing the equations in groups of two, beginning with the first two, a number of simpler relations are produced and by easy stages we obtain the results

$$A = -\frac{W}{12\beta}, \quad B = \frac{WL}{8\beta}\left(\frac{\mathscr{G}+2\beta}{\mathscr{G}+\beta}\right), \quad \phi = -\frac{WL^2}{2(\mathscr{G}+\beta)}, \quad M = 0; \\ \alpha = -\frac{WL}{4(\mathscr{G}+\beta)}, \quad \gamma = 0, \quad T = \frac{W\mathscr{G}L}{2(\mathscr{G}+\beta)}. \tag{4.2.6}$$

We are now able to examine the forces and displacements at all points of the structure. For example, the moment at 1 is given by

$$M_1 = 2B\beta = \frac{WL}{4}\left(\frac{\mathscr{G}+2\beta}{\mathscr{G}+\beta}\right). \tag{4.2.7}$$

The torque at the origin then becomes

$$T_1 = \mathscr{G}\alpha = -\frac{W\mathscr{G}L}{4(\mathscr{G}+\beta)}, \tag{4.2.8}$$

and the displacement under the load

$$\Delta_W = AL^3 + BL^2 = \frac{WL^3(\mathscr{G}+4\beta)}{24\beta(\mathscr{G}+\beta)}. \tag{4.2.9}$$

Shear at the origin is

$$F_1 = 6A\beta = -\frac{W}{2}.$$

The present structure clearly possesses various symmetries which have not been exploited. We have preferred to present the analysis in this way in order to display all the steps which would be necessary for a problem without symmetries. Finally we can view the results obtained and notice that for example the shear at 1 is $|W/2|$ as it should be from symmetry. Also $(T+M)_1 = -\frac{1}{2}WL$ as should be the case by taking moments about the axis 1–3 for the whole structure.

For many practical materials and section shapes the quantity $\mathscr{G} \doteqdot \beta$. Hence from (4.2.9) if $\mathscr{G} \doteqdot \beta$ we have

$$\Delta_W \doteqdot \frac{5}{48}\frac{WL^3}{\beta}. \tag{4.2.10}$$

If instead of the structure arranged as in fig. 4.2.0 we had arranged members 2–1, 2–3 to lie side by side to form a dual cantilever then

$$\Delta_W = \frac{8}{48}\frac{WL^3}{\beta}. \tag{4.2.11}$$

147

Hence we see that the L shape of structure is stiffer by some 40 % compared with an equivalent simple cantilever.

Now the moment under the load in our L shaped structure is

$$M_{2-} = (6AL + 2B)\text{ß}$$

$$= -\frac{WL\mathscr{G}}{4(\mathscr{G}+\text{ß})} \doteq -\frac{WL}{8} \quad (\text{if } \mathscr{G} \doteq \text{ß}). \qquad (4.2.12)$$

We have already seen that

$$M_1 = \frac{WL}{4}\left(\frac{\mathscr{G}+2\text{ß}}{\mathscr{G}+\text{ß}}\right) \doteq \frac{3WL}{8} \quad (\text{if } \mathscr{G} \doteq \text{ß}). \qquad (4.2.13)$$

These moments compare with the moment

$$M_1 = WL/2 \qquad (4.2.14)$$

for the double width cantilever.

Hence it is likely that the L member structure will be superior to the simple cantilever from both the stiffness and strength point of view although it should not be forgotten that M_1, M_2 in (4.2.13, 12) are associated with torques which are not present in the cantilever.

More extensive grillages of beams can be treated in like fashion to the above example.

4.3   Rigid-jointed space frameworks

Generally, in space frameworks with rigid joints, the members are bent about two axes as well as twisted. The methods which have been developed thus far for plane structures are easily extendable to space structures. But for the present we shall discuss a very simple configuration of space structure to illustrate certain other features of the discontinuity methods. The following example does not fully exploit the methods needed for space structures.

Consider the rigid-jointed framework shown in perspective in fig. 4.3.0. This frame has been analysed by a different method by Lightfoot and earlier by Rawlings (1.20). Provided we are concerned with the elastic response it is possible in this case to consider first the vertical plane frame $AEBC$ and the 10^u vertical point load together with a torsional restraint provided by BD; then the vertical plane frame CBD and the 1 unit|u distributed vertical load together with a torsional restraint provided by AEB. The frame is to be of common circular tubular section

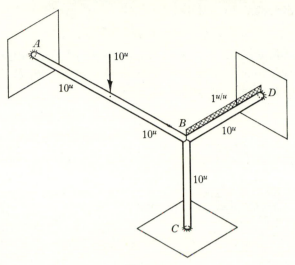

Fig. 4.3.0

throughout with $\mathscr{G} = 0.8\mathfrak{B}$. Consider first $AEBC$, fig. 4.3.1 (a), with the torsional restrain from BD of T as shown. Then

$$w = Ax^3 + Bx^2 + \frac{10z_e^3}{3!\,\mathfrak{B}} + \frac{Tz_b^2}{2!\,\mathfrak{B}} + \frac{Fz_b^3}{3!\,\mathfrak{B}} \quad (z_e = x - 10,\ z_b = x - 20). \quad (4.3.0)$$

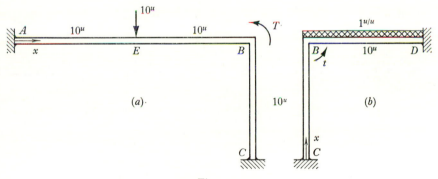

Fig. 4.3.1

This expression for the displaced form of $AEBC$ already contains the conditions at A, namely $w = w' = 0$ at $x = 0$. Also, no displacement discontinuity at D has been introduced in anticipation of D being a fixed point in space. There will however be a shear discontinuity at B, hence the $Fz_b^3/3!\,\mathfrak{B}$ term.

149

The conditions to be fulfilled are

$$w_D = 0; \quad w_D' = 10T/\mathscr{G},$$

the compatibility condition on torsion of BD; together with $w = w' = 0$ at C. There are four conditions and four unknowns A, B, T, F, hence we can solve. In turn the conditions are:

$$\left.\begin{array}{l}
8000A + 4000B + (10^4/6\beta) = 0, \\[4pt]
10T/\mathscr{G} = 1200A + 40B + 500/\beta, \\[4pt]
27000A + 900B + 8 \times 10^4/6\beta + 50T/\beta + 10^3F/6\beta = 0, \\[4pt]
2700A + 60B + 2 \times 10^3/\beta + 10T/\beta + 50F/\beta = 0.
\end{array}\right\} \quad (4.3.1)$$

Using $(4.3.1)_2$ to eliminate T we have

$$\left.\begin{array}{l}
318A + 10{\cdot}6B + \dfrac{5}{3}\dfrac{F}{\beta} + \dfrac{460}{3\beta} = 0, \\[10pt]
80A + 4B + 0 + \dfrac{50}{3\beta} = 0, \\[10pt]
366A + 9{\cdot}2B + \dfrac{5F}{\beta} + \dfrac{240}{\beta} = 0.
\end{array}\right\} \quad (4.3.2)$$

Now eliminating F from $(4.3.2)_{1,3}$ we obtain

$$588A + 22{\cdot}6B + 0 + (220/\beta) = 0,$$

whence $\qquad\qquad B\beta = 14{\cdot}3 \quad$ and $\quad A\beta = -0{\cdot}923. \qquad\qquad (4.3.3)$

Finally, $\qquad\qquad\quad T = -2{\cdot}95 \quad$ and $\quad F = -6{\cdot}70. \qquad\qquad (4.3.4)$

Here $A\beta$ and F have units of force; $B\beta$ and T, units of moment.

The second half problem is the analysis at CBD, fig. 4.3.1 (b). Now

$$w = ax^3 + bx^2 + \frac{tz_b^2}{2!\,\beta} + \frac{fz_b^3}{3!\,\beta} + \frac{z_b^4}{4!\,\beta} \quad (z_b = x - 10). \qquad (4.3.5)$$

In an exactly similar manner to the equations $(4.3.1)$ we now obtain

$$\left.\begin{array}{l}
1000a + 100b = 0, \\[8pt]
\dfrac{20t}{\mathscr{G}} = 300a + 20b, \\[10pt]
8000a + 400b + \dfrac{5t}{\beta} + \dfrac{10^3f}{6\beta} + \dfrac{10^4}{24\beta} = 0, \\[10pt]
1200a + 40b + \dfrac{10t}{\beta} + \dfrac{50f}{\beta} + \dfrac{10^3}{6\beta} = 0.
\end{array}\right\} \quad (4.3.6)$$

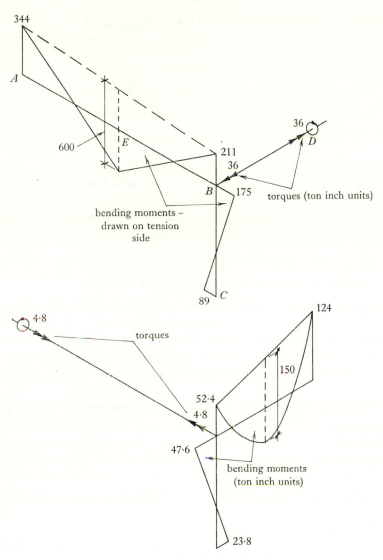

344

A

600

E

211

36

36

D

B 175·

torques (ton inch units)

bending moments –
drawn on tension
side

89 *C*

4·8

torques

124

150

52·4

4·8

47·6

bending moments
(ton inch units)

23·8

Fig. 4.3.2   Moments and torques in units of ton inches. Moments drawn on tension
face. Full displacement information is also available from (4.3.3, 4, 7).

Hence
$$a\beta = \tfrac{25}{252}, \quad b\beta = -\tfrac{250}{252}, \quad t = \tfrac{100}{252}, \quad f = -\tfrac{25}{3}. \tag{4.3.7}$$

Once again, $a\beta$ and f are forces; $b\beta$ and t are moments.

If now units of tons for force and feet for length are chosen then

$$M_A = 2B\beta = 28{\cdot}6\,T\,\text{ft} = 344\,T\,\text{inch},$$

$$M_C = 2b\beta = -1{\cdot}98\,T\,\text{ft} = -23{\cdot}8\,T\,\text{inch},$$

$$T = -2{\cdot}95\,T\,\text{ft} \doteqdot 36\,T\,\text{inch torque in } BD,$$

$$f = \tfrac{100}{252}\,T\,\text{ft} = 4{\cdot}8\,T\,\text{inch torque in } AB.$$

Proceeding thus the bending and twisting moments shown in fig. 4.3.2 are obtained. The diagrams are constructed with the moment ordinate drawn on the tension face of the beam.

We see that BA and BD are bent about a horizontal axis and twisted, while BC is bent about two horizontal axes.

4.4 Initial stress problems: I temperature stresses

Under this general head we group the effects resulting from internal stresses such as arise from temperature expansion, prestress from lack of fit either intentional or accidental and similar effects. The problems treated are entirely elementary and are intended merely to illustrate the method of approach. Elastic response is assumed.

To begin with, consider the frame of fig. 4.2.0 in the absence of external load but supposed subject to a uniform temperature rise, $J°$, the material of the frame being assumed to have a linear coefficient of expansion α.

Where in 4.2 the relevant displacements were w *transverse* to the plane of the frame and θ the beam twist, the situation in the present case is that all displacements will remain in plane and two components w *transverse* to the member but *inplane* and u the *axial* displacement, will be required to describe the deformation.

The G.D.E.s are
$$\left.\begin{aligned} \beta w^{\mathrm{iv}} &= 0, \\ u' &= \alpha J. \end{aligned}\right\} \tag{4.4.0}$$

Guided by our knowledge of discontinuous solutions, and with the same origin as in 4.2, the following general deflected shapes can be written down:

$$\left.\begin{aligned} w &= Ax^3 + Bx^2 + \Delta(z_2)° + (Fz_2^3/3!\,\beta), \\ u &= a + \alpha Jx + \delta(z_2)°. \end{aligned}\right\} \tag{4.4.1}$$

There are six unknowns A, B, Δ, F; a, δ which are found from the following conditions. The conditions on w at $x = 0$, $w = w' = 0$ are already incorporated in (4.4.1).

Now at $x = 0$, i.e. point 1, fig. 4.2.0,

$$u = 0 \quad \text{whence} \quad a = 0. \tag{4.4.2}$$

At the joint, point 2 $\qquad (w)_{2-} = (u)_{2+}$,

$$(w)_{2+} = -(u)_{2-},$$

whence
$$\left.\begin{array}{l} AL^2 + BL^2 = \alpha JL + \delta, \\ AL^2 + BL^2 + \Delta = -\alpha JL. \end{array}\right\} \tag{4.4.3}$$

At 3, $w = w' = u = 0$ whence

$$\left.\begin{array}{l} 8AL^3 + 4BL^2 + \Delta + (FL^3/6\beta) = 0, \\ 12AL^2 + 4BL + 0 + (FL^2/2\beta) = 0, \\ 2\alpha JL + \delta = 0. \end{array}\right\} \tag{4.4.4}$$

Hence
$$\left.\begin{array}{l} \Delta = 0, \quad AL^2 = 2\alpha J, \quad BL = -3\alpha J, \\ FL^2 = -24\alpha J\beta, \quad \delta = -2\alpha JL. \end{array}\right\} \tag{4.4.5}$$

From these results we can find the moments at 1 and 2 for example to be

$$\left.\begin{array}{l} M_1 = 2B\beta = -(6\alpha J\beta)/L, \\ M_2 = 6AL\beta + 2B\beta = (6\alpha J\beta)/L. \end{array}\right\} \tag{4.4.6}$$

Once again symmetry in the framework has not been utilized, in order to demonstrate the method the more clearly.

Exercise

Compute the moments at 1, 2, 3 in the unloaded frame fig. 4.2.0 if BC was too short by amount Ω during construction.

Ans.: $M_1 = 4 \cdot 5k$, $M_2 = -3k$, $M_3 = 1 \cdot 5k$; $k = \Omega\beta/L^2$.

4.5 Initial stress problems: II prestress

The use of prestress in concrete members to help counteract the effects of structural self weight is now a long established practice A typical situation is as illustrated diagrammatically in fig. 4.5.0.

The cable slope is everywhere small and the cable tension, T, is assumed constant throughout the beam. Hence the bending moment produced by the cable is at all points the local cable eccentricity from the line of centroids of the beam section multiplied by T, the cable tension. This moment will be denoted by small m. The sign convention for this moment will be as for the element fig. 1.1.0. This moment is the 'loading' on the beam. The *actual internal* beam moment can now be found by the usual integration procedures.

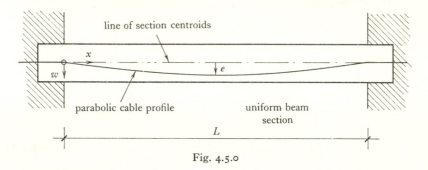

Fig. 4.5.0

For the parabolic profile shown in fig. 4.5.0 the m distribution will be

$$m(x) = \alpha x^2 + \beta x \tag{4.5.0}$$

for an origin as indicated. But $m'(L/2) = 0$ and $m(L) = 0$ from which, with $m(L/2) = Te$,

$$\alpha = -\frac{4Te}{L^2}, \quad \beta = \frac{4Te}{L}. \tag{4.5.1}$$

Hence
$$m(x) = \frac{4Tex}{L^2}(L-x). \tag{4.5.2}$$

If we denote the internal bending moment experienced by the *beam material* by M then we have the equilibrium equations

$$\left.\begin{aligned} F' &= p, \\ (M-m)' &= F, \end{aligned}\right\} \tag{4.5.3}$$

and the now familiar elastic beam material property equation

$$ßw'' = M. \tag{4.5.4}$$

In the present case, $p = 0$, hence
$$ßw^{\mathrm{iv}} = m'' = -\frac{8Te}{L^2}, \tag{4.5.5}$$

from which we see that the effect of the prestress is to give an *upward* uniformly distributed load of intensity $8Te/L^2$.

The solution now proceeds along the customary lines. The equation (4.5.5) solves to give

$$w = Ax^3 + Bx^2 + Cx + D - \frac{8Te}{L^2}\frac{x^4}{4!\,\text{ß}},$$ (4.5.6)

and the boundary conditions $w = w' = 0$ at $x = 0, L$ yield

$$D = C = 0; \quad B\text{ß} = -\frac{Te}{3}, \quad A\text{ß} = \frac{2}{3}\frac{Te}{L},$$ (4.5.7)

whence the check that $(SF)_0 = 4Te/L$ can be seen to apply

$$((SF)_0 = 6A\text{ß} = 4Te/L)$$

and

$$
\left.
\begin{aligned}
M(0) &= 2B\text{ß} = -\frac{2Te}{3}, \\
M(\tfrac{1}{2}L) &= (3AL + 2B)\,\text{ß} - Te = \frac{Te}{3}.
\end{aligned}
\right\}
$$ (4.5.8)

By suitable choice of T and e, a uniform beam weight together with added external u.d.l. can be made to produce a zero B.M. everywhere in the beam. The load is then carried by the cable tension and a uniform compression in the concrete. The axial shortening of the beam by virtue of this compression should not be forgotten about and may introduce residual effects in the remaining parts of the structure if the beam considered is part of a larger structure.

All that has previously been said about discontinuous solutions applies immediately to problems of prestressing cables such as that in this next example.

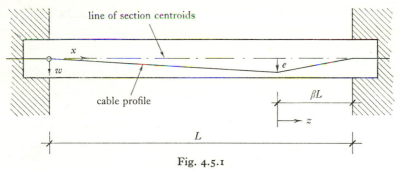

line of section centroids

x

w

cable profile

e

βL

z

L

Fig. 4.5.1

In the case of the prestressed beam fig. 4.5.1 we can write

$$w = Ax^3 + Bx^2 - \frac{Te}{\beta(1-\beta)L}\frac{z^3}{3!\,\text{ß}}, \quad z = x - L(1-\beta).$$

Whence $w = w' = 0$ at $x = L$, $z = \beta$ give

$$B\beta = -\frac{Te\beta}{2} \quad \text{and} \quad A\beta = \frac{Te\beta}{6L}\frac{3-2\beta}{1-\beta}$$

and
$$M(0) = 2B\beta = -Te\beta. \tag{4.5.9}$$

In both of the examples discussed, assuming the beam is first made with the cable unbonded and untensioned and then is finally (so-called) post-tensioned, the actual process of tensioning produces both internal bending moments and support reactions. This statement is more obviously true for a multi-span beam.

There are two particularly important classes of cable profile which have in turn the properties that they either produce bending moments but no reactions or, in contrast, produce reactions but no bending moments. The former class is the class of *concordant* profiles; the latter class that of *linear* profiles.

For a given beam to be prestressed there is nothing unique about a profile to satisfy the concordant requirement. In fact, any cable profile in the *shape* of a possible elastic bending moment distribution due to some arbitrary load system, but satisfying any required end or support conditions, will suffice.

The class of linear profiles is that in which the cable runs straight between support points, may have arbitrary eccentricity at a support or fixed end, passes through the centroidal point at a simple support and lies along the line of centroids in any cantilever section.

In the actual design of beams there is often the need to make changes to a cable profile because, for example, the eccentricities may be too great to be accommodated safely. A fruitful way in which to make changes is then by use of a superimposed linear profile since this does not alter the bending moments.

Essentially, to tackle any problem of the present type demands we establish or recognize the equivalent system of supporting loads produced by the action of tensioning the cable. The further analysis of a given situation is then routine but may include lack of fit terms of the type mentioned in 4.4 if the structure is redundant and post-tensioned.

4.6 Braced frameworks

The structures with which we have in the main been concerned are structures whose members are basically in bending, rather than in tension or compression as in a pin jointed truss. Typically they are rectangular beam and column constructions. Asymmetry of the structure or gross asymmetry in the loading applied to the structure cause the frame to sway, the joints displace and large moments are introduced into the members. Sway is to be avoided or counteracted as much as possible and one way of counteracting it is to brace the framework with light tension members across the diagonals of the rectangular spaces in the framework.

Clearly not all such openings can be braced without impairing the usefulness of the structure for access, placing of windows and so on and not all such openings need be braced. Typically one or a few bays in a multi-storey multi-bay frame can be braced with advantage. In this section we shall analyse the simplest such structure, the single bay, single storey cross braced portal.

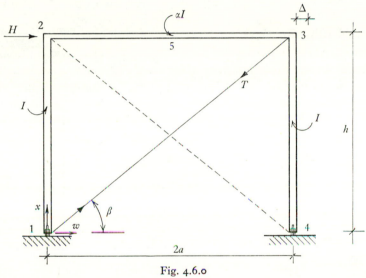

Fig. 4.6.0

Consider the rigid jointed frame shown in fig. 4.6.0 with a light tension bracing provided across 1–3 and 2–4. For the loading shown, the brace 1–3 is assumed to be just taut as the load H is applied. The brace 2–4 is assumed to become slack and can be neglected. The bracing is assumed to have a cross sectional area of A and Young's modulus E and only elastic effects will be considered. The aim is to establish the internal forces and displacements in terms of H.

Let the sway displacement be Δ. Then the bracing force T is given by

$$\frac{T}{AE}\frac{2a}{\cos\beta} = \Delta\cos\beta. \qquad (4.6.0)$$

The horizontal component of this force is the important quantity and is given by

$$T \cos \beta = \frac{AE \cos^3 \beta}{2a} \Delta = \Theta \Delta, \qquad (4.6.1)$$

where

$$\Theta = \frac{AE \cos^3 \beta}{2a}.$$

We shall now exploit the antisymmetry of the present framework which allows us to note that at the mid point of the beam, point 5,

$$w = w'' = 0, \quad x = h + a. \qquad (4.6.2)$$

Now the displaced form for a choice of origin at column foot 1 is given by

$$w = Ax^3 + Cx + \frac{Fz_2^3}{3!\,\alpha\beta} + \frac{Mz_2^2}{2!\,\alpha\beta} + \Delta_2, \qquad (4.6.3)$$

where the subscript $(\ )_2$ refers to point 2.

There are five unknowns in the displaced form. The five conditions are inextensibility of the column, moment continuity at 2, a shear balance for the beam and the two conditions (4.6.2). In the same order, these conditions are:

$$\left. \begin{aligned}
Ah^3 + Ch + \Delta_2 &= 0, \\
(1-\alpha)6Ah\beta &= M \quad (\beta = EI) \\
6A\beta + \tfrac{1}{2}H - \tfrac{1}{2}\Theta\Delta &= 0, \\
A(a+h)^3 + C(a+h) + \frac{Fa^3}{6\alpha\beta} + \frac{Ma^2}{2\alpha\beta} + \Delta_2 &= 0, \\
6A(a+h) + 0 + \frac{Fa}{\alpha\beta} + \frac{M}{\alpha\beta} + 0 &= 0.
\end{aligned} \right\} \qquad (4.6.4)$$

The sidesway displacement, Δ, in (4.6.4)$_3$ is given by

$$\Delta = Ah^3 + Ch. \qquad (4.6.5)$$

By easy steps the following results are obtained:

$$A = -\frac{\Delta}{2h^2(h + (a/\alpha))},$$

$$C = -h(3h + (2a/\alpha))\,A,$$

$$\Delta_2 = -\Delta,$$

whence in (4.6.4)$_3$ and using (4.6.5) we obtain

$$A = \frac{-H}{12\beta + 2h^2\Theta(h + (a/\alpha))}, \qquad (4.6.6)$$

and as a result the bending moment at 2 is given by

$$(M)_2 = 6A\beta h = -\frac{Hh}{2}\frac{1}{1+K},\qquad(4.6.7)$$

where

$$K = \frac{h^2\Theta(h+(a/\alpha))}{6\beta}.\qquad(4.6.8)$$

Finally

$$\Delta = -2h^2(h+(a/\alpha))\,A$$

$$= \frac{Hh^2(h+(a/\alpha))}{6\beta+h^2\Theta(h+(a/\alpha))}$$

$$= \frac{Hh^2(h+(a/\alpha))}{6\beta}\frac{1}{1+K},\qquad(4.6.9)$$

and K is as defined in (4.6.8). If $K = 0$ in (4.6.7, 9) we recover the knee moment and sway displacement for the unbraced frame.

The dimensionless number K is the significant parameter in the present study and is of the form bracing area $\times$ span^2/column stiffness. Written in another way, we can see that K is of the form

$$\frac{\text{bracing member area}}{\text{bending member area}} \times \left(\frac{\text{span of frame}}{\text{bending member radius of gyration}}\right)^2.\ (4.6.10)$$

Clearly this number, K, could be made quite large – of the order of ten quite easily. This would decimate the frame moments and sway displacement at the expense of adding the bracing member and some increase in the column loads. The effect of bracing can be therefore quite dramatic.

Exercise

For the same frame and choice of origin as in Fig. 4.6.0 but now with fixed column bases show that

$$w = Ax^3 + Bx^2 + \frac{Fz_2^3}{3!\,\alpha\beta} + \frac{Mz_2^2}{2!\,\alpha\beta} + \Delta_2$$

and

$$A = \frac{-2H(3h+(a/\alpha))}{24(3h+(a/\alpha))\beta+h^3\Theta(3h+(4a/\alpha))},$$

$$B = -1\cdot5hA\frac{3h+(2h/\alpha)}{3h+(a/\alpha)},$$

$$\Delta = \frac{(3h+(4a/\alpha))\,h^3H}{24(3h+(a/\alpha))\beta+h^3\Theta(3h+(4a/\alpha))},$$

$$\beta = (EI)_{\text{column}},\quad \Theta = (AE)_{\text{bracing}}\times(\cos^3\beta/2a).$$

Make estimates for the reduction in moments and displacements for feasible choices of bracing member sections.

4.7 Composite beams: shear connexion and slip

A class of beam which is becoming of increasing importance in some practical situations is the composite steel and concrete beam, where a concrete (compression) member is cast onto a rolled steel (tension) member and the interface suitably shear reinforced to ensure adequate transfer of shear between the two component parts of the composite section. The usual treatment of the interface is to butt weld stocky forged studs to the upper flange of the steel beam. These ensure good but not total bond between the steel beam and adjacent concrete and allow a small amount of slip.

We shall base our theory on the assumption that component parts of the beam themselves conform to beam bending theory and that at the interface between the two materials there is slip proportional to the interface shear stress. Consider the element of composite beam shown in Fig. 4.7.0.

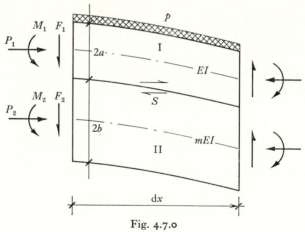

Fig. 4.7.0

In terms of the force resultants shown, the following equilibrium equations can be written down. For the upper member (I)

$$\left.\begin{aligned} M_1' - F_1 + P_1 w' - Sa &= 0, \\ P_1' &= S, \\ F_1' &= p - p_1, \end{aligned}\right\} \tag{4.7.0}$$

where p_1 is the interactive normal load intensity across the connexion interface and S the shear *force* per unit length across the interface.

For the lower member (II)

$$M_2' - F_2 + P_2 w' - Sb = 0,$$
$$P_2' = -S, \qquad (4.7.1)$$
$$F_2' = p_1.$$

The material property equations assumed will be the linear elastic equations

$$M_1 = \text{ß}w'',$$
$$M_2 = m\text{ß}w'', \qquad (4.7.2)$$

where $\qquad (EI)_{\text{I}} = \text{ß}, \quad (EI)_{\text{II}} = m\text{ß}.$

Thus far we have eight equations in $M_1, F_1, P_1; M_2, F_2, P_2; w, p_1, S$ – namely nine unknowns. The necessary equations are completed with the addition of the slip condition
$$(a+b)w' = \lambda S \qquad (4.7.3)$$

in which λ is a beam property and has dimensions of length/force. Axial extension of either component member along its centroidal line is assumed zero in writing (4.7.3) down.

Let us now derive the governing equation for this problem. By taking sums of appropriate equations between (4.7.0–1) we readily obtain

$$(1+m)\,\text{ß}w^{\text{iv}} - [(a+b)^2/\lambda]\,w'' = p$$

or
$$w^{\text{iv}} - \frac{(a+b)^2}{\lambda\text{ß}(1+m)}\,w'' = \frac{p}{\text{ß}(1+m)}. \qquad (4.7.4)$$

This is the governing equation for the composite beam. Once w is known S, M_1, M_2 can be found. However P_1, P_2, F_1, F_2 can only be found to within additive constants, since

$$P_1 = \frac{a+b}{\lambda}w + c, \quad P_2 = -\frac{a+b}{\lambda}w + d.$$

If
$$\beta^2 \equiv \frac{(a+b)^2}{\lambda\text{ß}(1+m)}, \qquad (4.7.5)$$

then the general solution of (4.7.4) becomes
$$w = A\sinh\beta x + B\cosh\beta x + Cx + D - \frac{\lambda p x^2}{2(a+b)^2}. \qquad (4.7.6)$$

The particular integral included here assumes p is a constant distributed load.

The equation (4.7.4) is of fourth order and hence we are at liberty to specify only two boundary conditions at each end of a beam although more seem to be needed physically.

For example, a simply supported end requires that

$$w = 0, \quad (M_1 + M_2) = 0, \tag{4.7.7}$$

with $P_2 = P_1 = 0$ at the same point. This latter condition supplies the additive constants referred to above as zeros. The true boundary conditions are (4.7.7).

Likewise fixity means $w = w' = 0$ and a free-end requires $M_1 + M_2 = 0$, $F_1 + F_2 = 0$ (with $P_1 = P_2 = 0$). These conditions require that:

$$
\left.
\begin{aligned}
F_1 + F_2 &= (M_1 + M_2)' - (a+b)\,S \\
&= (1+m)\,\beta w''' - \frac{(a+b)^2}{\lambda}\,w' = 0, \\
M_1 + M_2 &= (1+m)\,\beta w'' = 0.
\end{aligned}
\right\} \tag{4.7.8}
$$

Hence the most common boundary conditions can be summarized as

$$
\left.
\begin{aligned}
\text{fixity} &\qquad w = w' = 0, \\
\text{simple support} &\qquad w = w'' = 0, \\
\text{free-end} &\qquad w'' = w''' - \beta^2 w' = 0.
\end{aligned}
\right\} \tag{4.7.9}
$$

Note

There may be some situations in which the inextensibility of each element of the composite is not tenable. Then if we allow for axial compression, all the equations (4.7.0, 1–2) are unchanged but the slip condition becomes

$$(a+b)\,w' - \int_0^x \frac{P}{AE}\left(1 + \frac{1}{n}\right) dx = \lambda s, \tag{4.7.10}$$

where now we have AE values for I and II in the ratio $1:n$ and $P = P_2 = -P_1$. This latter condition will hold if there is a free end or simple support in the beam. The origin for x in this expression is to be a point where there is zero slip. The best choice is a line of symmetry for the beam and loading.

The equations (4.7.0, 1, 2, 10) can now be seen to be consistent with the homogeneous beam equations when the elements are identical. For them $m = n = 1$, $a = b$ and $\lambda = 0$.

Hence, from (4.7.10), $\qquad P_2 = AEaw''.$

Now the resultant moment $\mathcal{M}$ across any section is given by

$$
\begin{aligned}
\mathcal{M} &= M_1 + M_2 + 2P_2 a \\
&= (2\beta + 2a^2 AE)\,w'' \\
&= 2E(I_{11} + Aa^2)\,w'' \\
&= \beta w'', \tag{4.7.11}
\end{aligned}
$$

where ß is the *EI* of the whole section about the axis *O–O*. The shear force *S* is then an *internal* stress resultant and cannot be found from (4.7.0, 1 and 11).

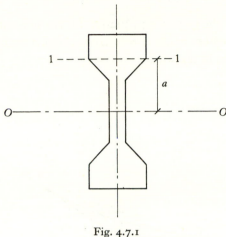

Fig. 4.7.1

In (2.24) we make mention of the analogy between the present problem and the multi-storey frame problem studied in chapter 2, part II. There are many crosslinks which unfortunately we are unable to elaborate upon here.

4.8   The cable: small dip, inextensional theory

Thus far all the structural elements we have considered have been bent and have offered resistance to this bending. A cable is a structural element which, although bent, offers *no* resistance to bending. The cable supports self weight and superimposed loading through the tension developed. The shape of the cable accommodates the loads acting so as to ensure that the cable tension is always tangential to the cable profile. A typical element of cable is shown in fig. 4.8.0. Here ψ is assumed small; then horizontal equilibrium ensures that T, the cable tension, remains constant – in the absence of any applied horizontal load. Vertical equilibrium gives

$$T\tan(\psi+\mathrm{d}\psi) - T\tan\psi = p\,\mathrm{d}s.$$

For $\psi \ll 1$ then $\qquad \mathrm{d}s \doteqdot \mathrm{d}x, \quad \tan\psi \doteqdot \psi.$

But $\qquad\qquad\qquad \psi = \dfrac{\mathrm{d}w}{\mathrm{d}x}.$

Hence
$$T(\psi + \mathrm{d}\psi) - T\psi = p\,\mathrm{d}x,$$

$$T\frac{\mathrm{d}\psi}{\mathrm{d}x} = p \quad \text{or} \quad T\frac{\mathrm{d}^2 w}{\mathrm{d}x^2} = p. \qquad (4.8.0)$$

Equation 4.8.0 is the governing equation for the present study of cables. Note the displacement w direction is here the reverse of the convention adopted elsewhere in the work.

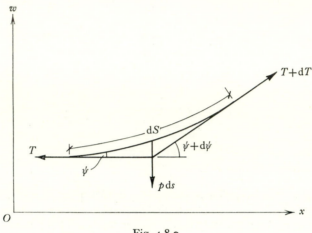

Fig. 4.8.0

Because $T = \text{constant}$, (4.8.0) can be immediately integrated to give

$$w = Ax + B + (px^2/2T). \qquad (4.8.1)$$

Here A and B are respectively, the slope and displacement of the cable at the chosen origin. The particular integral shown assumes $p = \text{constant}$.

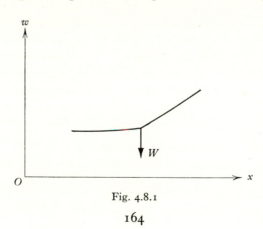

Fig. 4.8.1

4.8 Cables: small dip, inextensional theory

As in all the preceding discussion, so too here, considerations of discontinuity play an important part in the solution of physically interesting problems. A discontinuity in a physical variable of interest in practice is the concentrated gravity load applied to the cable. Such a discontinuity in the shear on the cable is sustained by the cable tension acting across a *slope discontinuity*. Thus, for load W and cable tension T,

$$[\![dw/dx]\!] = W/T.$$

Another requirement is often to evaluate the actual length of the cable. Thus

$$ds^2 = dx^2 + dw^2,$$

$$ds = \sqrt{(1 + (dw/dx)^2)}\, dx \doteq (1 + \tfrac{1}{2}(dw/dx)^2)\, dx \quad (dw/dx \ll 1),$$

or

$$s = \int (1 + \tfrac{1}{2}(dw/dx)^2)\, dx. \tag{4.8.2}$$

For a symmetrical cable, spanning L, with a central dip d and under action of the self weight p/unit length only, then, with an origin attached to the cable at the lowest point, we have

$$A = B = 0 \quad \text{whence} \quad dw/dx = px/T,$$

and

$$s = 2 \int_0^{\frac{1}{2}L} \left(1 + \frac{1}{2} \left(\frac{px}{T} \right)^2 \right) dx = L + \left(\frac{p}{T} \right)^2 \left(\frac{L^3}{24} \right).$$

But from (4.8.1) with $x = \tfrac{1}{2}L$,

$$w(\tfrac{1}{2}L) = d = (p/2T)(\tfrac{1}{2}L)^2 \quad \text{or} \quad (p/T) = 8d/L^2.$$

Hence

$$s = L(1 + \tfrac{8}{3}(d/L)^2). \tag{4.8.3}$$

Suppose a cable of weight 1.5 kgf/m spans 1600 m with a 2000 kgf concentrated load at midspan and a central dip of 20 m. Then, with the choice of origin as shown,

$$B = 0, \quad A = \text{slope at origin}$$

$$= \left(\frac{\text{shear force}}{T} \right)_{\text{origin}}$$

$$= \frac{1000}{T}, \quad T = \text{cable tension}.$$

Hence

$$w_2 = 20 = \frac{1000}{T} 800 + \frac{1.5(800)^2}{2T}, \quad T = 64{,}000 \text{ kgf}.$$

165

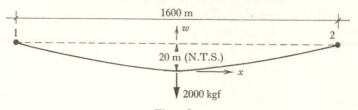

Fig. 4.8.2

The dip at quarter span is given by

$$(\text{Dip})_{400\,\text{m}} = 20 - \left(\frac{400}{64} + \frac{1 \cdot 5 \times 400^2}{2 \times 64\,000}\right)$$

$$= 20 - 8 \cdot 12 = 11 \cdot 88\,\text{m}. \qquad (4.8.4)$$

The actual length of the cable, assumed inextensible, is given by

$$s = 2\int_0^{800} \left(1 + \frac{1}{2}\left(\frac{dw}{dx}\right)^2\right)dx = 1600 + \int_0^{800}\left(A + \frac{px}{T}\right)^2 dx$$

$$= 1600 + 0 \cdot 523\,\text{m}.$$

Suppose the 2000 kgf load is transferred to the quarter span position. What now is the cable tension and the dip at the load point?

The origin will be chosen at the end as indicated in fig. 4.8.3. Then

$$w = Ax + B + (px^2/2T) + (2000/T)z,$$

$$z = x - 1200.$$

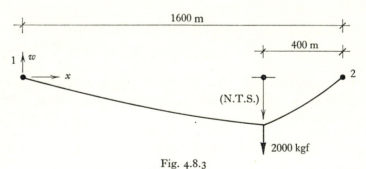

Fig. 4.8.3

Again $B = 0$, because the origin is a point on the cable. The condition that end 2 is level with 1 is given by

$$0 = A \times 1600 + \frac{1 \cdot 5 \times 1600^2}{2T} + \frac{2000}{T}\,400$$

or

$$A = -1700/T. \qquad (4.8.5)$$

4.8 Cables: small dip, inextensional theory

The length of cable is unchanged, hence

$$1600 + 0{\cdot}523 = 1600 + \frac{1}{2}\int_0^{1600}\left(A + \frac{px}{T}\right)^2 dx$$

$$+ \int_{1200}^{1600}\left(A + \frac{px}{T}\right)\left(\frac{2000}{T}\right) dx + \frac{1}{2}\int_0^{400}\left(\frac{2000}{T}\right)^2 dz,$$

or

$$0{\cdot}523 = \frac{1706 \times 10^6}{T^2},$$

whence

$$T = 57\,200\ \text{kgf.} \tag{4.8.6}$$

Finally, the dip at the load point equals

$$-A \times 1200 - \frac{1{\cdot}5 \times 1200^2}{2T} = \frac{1200}{57\,200}\left(1700 - \frac{1{\cdot}5 \times 1200}{2}\right)$$

$$= 16{\cdot}75\ \text{m.} \tag{4.8.7}$$

This example illustrates most of the features of such problems. More intricate problems, which lack any symmetry, with further loads, and so on, can be treated in like fashion.

4.9 Conclusions and a look ahead

In this chapter we have collected together brief discussions of a number of topics which have in common only the one dimensional character of the underlying differential equations. So far, of the entire field of one dimensional structures, the area not yet touched upon is the field of structures with *curved* members. This subject it is planned to discuss in a companion volume.

Again, little has been said about computation using a digital computer, although our methods are well adapted to such computation. This topic too is deferred for later discussion. Efficient machine computation requires extensive use of matrices. But although the transition to a matrix formulation can be logically accomplished, the matrices themselves are an unnecessary and cumbersome tool for *manual* computation. This is the reason for avoiding all use of matrices in the present treatment, since the subject is here really being discussed from the view point of use in manual computation.

As has been hinted at in the preface, a major modern interest in structural mechanics centres around two dimensional plate and shell structures. In this area of the subject, much of our discussion of method in the preceding chapters finds a natural place although the need for mechanized computation becomes more acute. We trust that the reader's interest will carry him into this field.

Apart from the novelty of actually formulating the two dimensional problems there remains the even bigger field of problem solution. For the engineer this means the need to employ approximate methods of one sort or another. In the next chapter we discuss one approximation procedure which is useful also in the two dimensional field.

4.10   References

Guyon, Y. *Prestressed Concrete*, C.R. Books (1963). A translation of a standard French work by a former associate of the late Eugène Freyssinet.

Hendry, A. W. and Jaeger, L. G. *The Analysis of Grid Frameworks and Related Structures*, Chatto and Windus, London (1958). Although there is no adequate account of grillage problems in the literature this book goes some way to meeting the needs.

Matheson, J. A. L. *Hyperstatic Structures*, 2 vols., Butterworth, London (1959). We recommend these books, especially volume II, as a source of useful problems with answers.

Most of the texts cited in the references to chapters 1–3 discuss temperature stress problems and the interested reader can usefully reformulate problems from these sources using the present approach.

5

FOUNDATIONS OF BEAM THEORY
AND APPROXIMATIONS

5.0   Introduction

In the foregoing chapters we have given a development of our subject
which emphasizes the underlying differential nature of the problems
discussed. For the scheme to be complete we must discuss means for
deducing the differential equations in new situations and also indicate
how problems may be solved approximately if the time available or
equation difficulty preclude a complete solution. These two aspects will
be discussed in the present chapter, although only briefly. The methods
used are applicable to multi-dimensional situations but the problems we
shall deal with will relate specifically to one dimension. We shall also be
dealing with rectangular Cartesian coordinate systems exclusively. Curvilinear systems can be included with appropriate generalizations of the
tensor quantities involved.

Before beginning the discussion of the foundations of beam theory it
is necessary to set down some facts concerning functions of one or more
variables and their characterization.

5.1   Functions and moments

Once we wish to deal with functions more complicated than simple
constants or linear variations, various methods for describing the functional variation are possible. We have been concerned in the foregoing
chapters with explicit solution of given ordinary differential equations
and the problem of functional description has not intruded. In wishing
to discuss the *Foundations* of a theory we are really attempting to view the
given theory as an 'approximation' to some wider and more general
theory and there is then an immediate need to review what we mean by
an 'approximation'.

Basically, approximation schemes relate directly to functional approximations and in turn to the description of functional variations. We
propose to exploit the notion of *moments* of a function as a tool in such
a description.

The *moments* of a function $f(z)*$ on the domain $-h \leqslant z \leqslant h$ are the quantities $f^{(n)}$ defined by

$$f^{(n)} \equiv \frac{1}{2h} \int_{-h}^{h} f(z) \, z^n \, \mathrm{d}z \quad (n = 0, 1, 2, \ldots).\dagger \qquad (5.1.0)$$

We see that the $f^{(n)}$ contain no explicit z variation but are infinite in number where $f(z)$ was a single function with explicit z variation. Now there is a theorem in mathematical analysis which tells us that knowledge of the $f^{(n)}$ for all n yields full information about $f(z)$, although the construction of the $f(z)$ from the $f^{(n)}$ may not be easy. In a practical case we have access to only a finite number of $f^{(n)}$ and herein lies our approximation.

The expression (5.1.0) may seem a curious definition to wish to invoke, until we recognize that at many points in engineering computation and design we are in fact dealing with $f^{(n)}$ and sometimes we even infer – often by means of added understanding of the problem – the nature of the $f(z)$ variation itself. Thus if $f(z)$ is a stress variation, $f^{(0)}$ is the *resultant force* per unit length across a $2h$ deep strip and $f^{(1)}$ is the *resultant moment* per unit length (except for a non-essential factor of $2h$ compared with the usual definition). Or again, for an even function and $h \to \infty$, $f^{(2n+1)} = 0$ and $f^{(0)}$ is a measure of the mean and $(f^{(2)})^{\frac{1}{2}}$ of the standard deviation for a probability density $f(z)$. The concept of moment is a deeply engrained concept and can usefully form the basis for a systematic theory of approximation. This is what we propose to do although it should be added that we have no immediate backing from mathematical theory for the proposals we are about to make. They are at best engineering extrapolations from experience.

In the application of *moments* in probability theory, the lowest moments, usually zeroth and second, are used to characterize the density function. This is a complete description since the density function is known or assumed, the range of integration is infinite and of all the moments only two are independent.

In the present application of *moment* concepts, the central feature of the problems is that just to take the moments of the original equations produces more unknowns than equations and hence some *approximation* must be made in order to obtain a soluble system.

---

* The z used here is not the z of earlier chapters. The present z is an independent real variable and one of the trio of coordinates x, y, z.
† The $(\ldots)^{(n)}$ notation here is different from the same notation defined on p. 1.

5.2 Some operations with moments

We have already remarked that our entire discussion will be limited to rectangular Cartesian frames. The following definitions and deductions will be used later.

Given an $f(z)$, which may also be a function of x and y, and an interval $-h \leqslant z \leqslant h$, then we define

$$f_{(\pm)}(z) \equiv \tfrac{1}{2}(f(h) \pm f(-h)). \tag{5.2.0}$$

We shall require an expression for $\left(\dfrac{df(z)}{dz}\right)^{(n)}$. Now, integrating by parts, we have

$$\left(\frac{df(z)}{dz}\right)^{(n)} \equiv (f(z),_3)^{(n)} \equiv \frac{1}{2h}\int_{-h}^{h} \frac{df(z)}{dz} z^n \, dz$$

$$= \frac{1}{2h}\left(z^n f(z)\Big|_{-h}^{h} - \int_{-h}^{h} f(z)\, nz^{n-1} \, dz\right)$$

$$= h^{n-1} f_{(\mp)}(z) - nf^{(n-1)}, \tag{5.2.1}$$

where the $(\mp)$ sign is chosen according as n is $-$ even, $+$ odd, respectively. As convenient, we shall regard the x, y, z coordinate system as x_1, x_2, x_3 and abbreviate them to x_i ($i = 1, 2, 3$). If there is need to refer to x_1, x_2 alone, the notation x_α ($\alpha = 1, 2$) will be adopted, using a Greek rather than a Latin suffix.

5.3 The approximation procedure

Our empirical extrapolation from experience which will enable us to use moments as a viable approximation device is to say that *more information is conveyed by the lower than the higher moments of a function*. There is no firm mathematical basis for this assertion but provided $f(z)$ is not too wildly varying then the statement provides a useful basis for approximation.

Again, we have defined a moment in terms of a single variable z, but multi-dimensional moments can also be defined and used to advantage. An example is an $f(x, y)$ giving rise to

$$f^{(m, n)} \equiv \int_{-h}^{h}\int_{-k}^{k} f(x, y)\, x^n y^n \, dx \, dy. \tag{5.3.0}$$

A two-dimensional moment of physical occurrence is $f^{(0, 0)}$, where f is the warping function of the elastic torsion problem and $f^{(0, 0)}$ is then proportional to the torsion constant; also second moments of area, I_{ij}.

5.4 The elasticity equations

The topic 'foundations of beam theory' we shall take to mean the study of the relationship between the technical theory already discussed at length and the deduction of this and other theories from the underlying three dimensional elastic equations. Thus, we shall aim to deduce from the three dimensional elastic equations, the technical theory and show how more sophisticated and comprehensive one dimensional theories may be derived, especially the so-called deep-beam theory, the relevant theory when shear deformations, as well as those deriving from bending deformation, become significant.

To begin with though, let us familiarize ourselves with the elasticity equations. The Cartesian stress tensor will be denoted by t_{ij} and the Cartesian displacement vector by u_i. Then, if the coordinate axes are principal stress axes, the principal stresses will be t_{11}, t_{22}, t_{33}, tensile positive, acting on planes normal to the x, y, z axes respectively. The shear stresses t_{12}, t_{23}, t_{31} will then be zero. Likewise u_i is the vector with components (u_1, u_2, u_3) being the displacements along the x, y, z coordinate directions respectively.

Then in rectangular coordinates the equations of classical linear elasticity for an isotropic solid are the three equilibrium equations

$$t_{ij,j} + \mathscr{F}_i = 0, \qquad (5.4.0)$$

together with the six stress–displacement equations

$$t_{ij} = \mu(u_{i,j} + u_{j,i}) + \lambda \delta_{ij} u_{k,k}. \qquad (5.4.1)$$

In (5.4.0, 1) the comma $(,_j)$ denotes differentiation with respect to x_j; μ, λ are the Lamé elastic constants; δ_{ij} is the Kronecker delta tensor and the summation convention is employed. In addition, the stress tensor is symmetric since couple stress resultants and body couple are not admitted and $\mathscr{F}_i$ represents body forces and reversed mass–acceleration terms. The equations (5.4.0, 1) are nine equations in three independent variables and in the nine unknowns u_i, t_{ij}.

We do not propose to give a discussion of Cartesian tensors. The interested reader should refer to Jeffries, Temple or Pearson (5.9).

5.5 The foundations of beam theory

In the previous section we have set down the elastic continua equations as they apply in three dimensions. We wish to examine the relationship between these equations and possible one dimensional beam theories, in particular the Euler–Bernoulli theory used throughout the earlier chapters, and to develop a more complete theory which can be termed a deep-beam theory for use in situations where deformations in addition to the bending deformations of Euler–Bernoulli theory are significant.

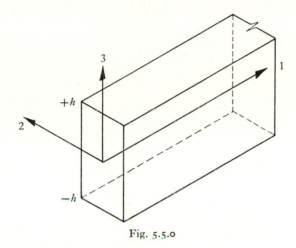

Fig. 5.5.0

Our method of approach is to take the moments of the nine elastic equations (5.4.0, 1) with respect to the coordinate through the *depth* of the beam section (the axis 3 in fig. 5.5.0). We shall find it convenient to give the coordinate $3(z)$ prominence and denote 1, 2 collectively by α. Then the moments of (5.4.0, 1) can be written as

$$\left.\begin{array}{l} t_{\alpha 3,\,\alpha}^{(n)}+h^{n-1}t_{(\pm)33}-nt_{33}^{(n-1)}+\mathscr{F}_3^{(n)}=0,\\[2mm] t_{\alpha 3}^{(n)}=t_{3\alpha}^{(n)}=\mu\big(h^{n-1}u_{(\pm)\alpha}-nu_\alpha^{(n-1)}+u_{3,\,\alpha}^{(n)}\big), \end{array}\right\} \qquad (5.5.0)$$

and

$$\left.\begin{array}{l} t_{\alpha\beta,\,\alpha}^{(n)}+h^{n-1}t_{(\pm)3\beta}-nt_{3\beta}^{(n-1)}+\mathscr{F}_\beta^{(n)}=0,\\[2mm] t_{\alpha\beta}^{(n)}=\mu\big(u_{\alpha,\,\beta}^{(n)}+u_{\beta,\,\alpha}^{(n)}\big)+\lambda\delta_{\alpha\beta}\big(u_{\tau,\,\tau}^{(n)}+h^{n-1}u_{(\pm)3}-nu_3^{(n-1)}\big),\\[2mm] t_{33}^{(n)}=(\lambda+2\mu)\big(h^{n-1}u_{(\pm)3}-nu_3^{(n-1)}\big)+\lambda u_{\tau,\,\tau}^{(n)}. \end{array}\right\} \qquad (5.5.1)$$

Here $n=0,1,2,\ldots$ and the $(+)$ $(-)$ sign is chosen according as n is $+$ odd, $-$ even. The equations (5.5.0, 1) look rather complicated. We observe however that they constitute two separate groups of equations.

173

Thus, (5.5.0) for n even taken with (5.5.1) for n odd on the one hand, and (5.5.0), n odd, with (5.5.1), n even, on the other, constitute two separate groups of equations. The former describe *bending* and the latter *stretching* of the elastic material. In our discussion here we shall examine only the bending group.

Now write down the equations (5.5.0) for $n = 0$ and (5.5.1) for $n = 1$. Then with no body force or mass acceleration terms, $\mathscr{F}_i = 0$ and

$$\left.\begin{aligned}
t^{(0)}_{\alpha 3, \alpha} &+ (1/h)\, t_{(-)33} = 0, \\
t^{(0)}_{\alpha 3} = t^{(0)}_{3\alpha} &= \mu((1/h)\, u_{(+)\alpha} + u^{(0)}_{3,\alpha}), \\
t^{(1)}_{\alpha\beta, \alpha} &+ t_{(+)3\beta} - t^{(0)}_{3\beta} = 0, \\
t^{(1)}_{\alpha\beta} = \mu(u^{(1)}_{\alpha,\beta} &+ u^{(1)}_{\beta,\alpha}) + \lambda\delta_{\alpha\beta}(u^{(1)}_{\tau,\tau} + u_{(+)3} - u^{(0)}_3), \\
t^{(1)}_{33} &= (\lambda + 2\mu)((u_{(+)3} - u^{(0)}_3) + \lambda u^{(1)}_{\tau,\tau}.
\end{aligned}\right\} \quad (5.5.2)$$

The equations (5.5.2) can be further simplified by noting that $(..)_{,2} = 0$ in all our beam problems. This is a practical simplification. Hence

$$u^{(0)}_{\tau,\tau} = u^{(0)}_{1,1}. \quad (5.5.3)$$

In bending problems u_3, $t_{\alpha\beta}$ and $t_{\alpha 3}$ will be even functions of coordinate 3 and u_α, t_{33} will be odd functions. Physically t_{33} is the least important of the stresses and it is probably permissible to neglect $t^{(1)}_{33}$.

Hence from $(5.5.2)_5$

$$u^{(1)}_{1,1} = \frac{\lambda + 2\mu}{\lambda}(u^{(0)}_3 - u_{(+)3}). \quad (5.5.4)$$

Of the displacements, $u_{2,2} = 0$ and u_1 is odd in x_3. The simplest variation of u_1 is linear whence, if $u = u(x_1)$,

$$u_1 \equiv -u \cdot x_3.$$

As a result $\qquad u_{(+)1} = -\tfrac{1}{2}u\,[h - (-h)] = -uh,$

and $\qquad \begin{aligned} u^{(1)}_1 &= -\frac{1}{2h}\int_{-h}^{h} u x_3^2 \, dx_3 \\ &= -\frac{uh^2}{3} = -\frac{u_{(+)1} \cdot h}{3}. \end{aligned} \quad (5.5.5)$

Now in $(5.5.2)_4$ for $\alpha = \beta = 1$,

$$\begin{aligned}
t^{(1)}_{11} &= (\lambda + 2\mu)\, u^{(1)}_{1,1} - \frac{\lambda^2}{\lambda + 2\mu}\, u^{(1)}_{1,1} \\
&= \frac{4\mu(\lambda + \mu)}{(\lambda + 2\mu)}\, u^{(1)}_{1,1} = -\frac{4}{3}\frac{\mu(\lambda + \mu)}{(\lambda + 2\mu)}\, u_{,1} h^2.
\end{aligned} \quad (5.5.6)$$

From $(5.5.2)_3$, if $t_{(+)3\beta} = 0$,

$$t^{(1)}_{1\beta,1} - t^{(0)}_{3\beta} = 0,$$

and, from $(5.5.2)_1$, $\qquad t^{(0)}_{13,1} + (1/h)\, t_{(-)33} = 0.$ $\qquad (5.5.7)$

The $t_{(+)3\beta} = 0$ condition merely says that there is no surface shear loading. This is true of most practical situations, although not for the composites of a composite beam.

We are now in a position to redefine the moments occurring in terms of more familiar resultants from beam theory. Thus

$$2hbt^{(0)}_{13} \equiv -F,$$
$$u^{(0)}_3 \doteq u_{(+)3} \equiv w,$$
$$2hbt^{(1)}_{11} \equiv -M,$$
$$2bt_{(-)33} \equiv p,$$

$\qquad (5.5.8)$

where F, w, M, p have the meanings assigned to them in 1.1, 1.7.

Then $(5.5.7)$, $(5.5.6)$, $(5.5.2)_2$ become

$$M' = F, \quad (\ )' = \mathrm{d}(\ )/\mathrm{d}x, \quad x = x_1$$
$$F' = p,$$
$$M = \frac{\frac{8}{3}bh^3\mu(\lambda+\mu)}{\lambda+2\mu}u' \equiv \beta u',$$
$$F = 2hb\mu(u-w') \equiv \mathscr{D}(u-w').$$

$\qquad (5.5.9)$

Now if $\qquad \lambda \equiv \dfrac{E^*\nu}{(1+\nu)(1-2\nu)} \quad \text{and} \quad \mu \equiv \dfrac{E^*}{2(1+\nu)},$

defining E^* and ν; then

$$\beta = \frac{E^*I}{1-\nu^2}, \quad \mathscr{D} = \mu A \quad \text{with} \quad A = 2bh, \quad I = \frac{2bh^3}{3}.$$

A is the section area and I the centroidal second moment of area.

The group of four equations $(5.5.9)$ constitute a suitable *deep-beam theory*. If, however, the final equation, $(5.5.9)_4$, is written

$$F = 0 = \mathscr{D}(u-w') \qquad (5.5.10)$$

then we recover the Euler–Bernoulli theory.

This latter equation is essentially the requirement that the originally plane sections of the beam normal to the beam axis x_1 remain plane and *normal* to the beam axis when the beam is bent. The deep-beam theory as we have developed it in (5.5.9) dispenses with the *normal* condition but retains the plane condition.

To bring the present theory into line with other developments found in the literature, we should note that the composite elastic constant $E^*/(1-\nu^2)$ will be called Young's modulus for beams and will be denoted by E. The shear modulus G will be given by $G = \mu/(1-\nu^2)$.

Thus, within a very brief compass, we have been able, by use of moment concepts, to give a derivation of beam theory suitable for all practical purposes. The real importance of the moment method is however not in allowing us to make the present deductions but in providing a systematic manner in which yet more refined theories can be derived by writing down more moment equations, should the need arise.

And especially, too, moment methods provide a means to obtain approximate solutions to given problems described by differential equations. This latter feature will be illustrated with an example later in the chapter. Next we shall examine the deep-beam theory just derived by means of an example.

Note

5.1 We have indicated one method for examining the relation between one, two and three dimensional theories. This is not the only method available, indeed it is not the method generally used. The most generally used approach is via an energy argument and the reader is referred to Washizu (5.9) for such a discussion. A point of interest however is that the theories so obtained are in all important respects similar to but not identical with that derived above. Thus in the case of a rectangular beam, the elastic constant $\mathscr{D}$ takes the value $\frac{5}{6}GA$ in the usual energy derivation whereas here it takes the value $G(1-\nu^2)A$. Since the effect being described by the $\mathscr{D}$ multiplied terms is of second order, the differences thus introduced are of negligible effect in practice.

5.6   A second order beam theory: shear deformation

Thus far our entire discussion of beams, frames and columns has been based on the single moment/displacement relation

$$M = \text{ß}w''. \tag{5.6.0}$$

Provided that the span (L)/depth of section (D) ratio is of the order of 20 or greater this relation produces a viable theory which can be checked against experiment satisfactorily.

If however the beam section is relatively deeper so as to reduce this L/D ratio, then not all the deformation can be accounted for from bending

5.6 *Second order beam theory: shear deformation*

– there is some additional deformation arising from shear distortion. The 'plane sections remaining plane *and normal*' requirement, which is built into (5.6.0) must now be changed and a 'better' theory used. It has been shown in the previous section that a suitable new theory is obtained by replacing (5.6.0) with the following pair of force–displacement relations:

$$M = ßu', \\ F = \mathscr{D}(u - w'). \quad (5.6.1)$$

Here M, F, w have the meanings previously assigned to them and u is a new (non-dimensional) variable related to the distortion of the plane cross section with respect to the centre line. We shall leave it as an exercise to the reader to explore the significance of u. The elastic constant $\mathscr{D}$ has the value $G(1 - \nu^2)A$ for a rectangular section. Here A is the section area, G the shear modulus and ν Poisson's ratio.

If equations (5.6.1) are now allied with the familiar equilibrium equations for the beam element, namely

$$M' = F, \\ F' = p, \quad (5.6.2)$$

then it is easy to see that

$$M = ßu' = ß\left(w' + \frac{F}{\mathscr{D}}\right)' = ßw'' + \frac{ß}{\mathscr{D}}F' \\ = ßw'' + (ß/\mathscr{D})p, \\ F = ßw''' + (ß/\mathscr{D})p', \\ u = w' + (ß/\mathscr{D})w''' + (ß/\mathscr{D}^2)p', \quad (5.6.3)$$

and finally that the G.D.E. is

$$ßw^{iv} = p - (ß/\mathscr{D})p''. \quad (5.6.4)$$

The new beam theory is still a very simple theory as evidenced by the G.D.E. being fully integrable in quadratures, but interesting and useful additional effects can now be studied.

First, the general solution is

$$w = Ax^3 + Bx^2 + Cx + D + \text{P.I.} \quad (5.6.5)$$

as for the first order theory. Because the order of the G.D.E. has not changed, the same number of boundary conditions, at two per end, still apply.

Particular integrals, and the boundary conditions themselves, are however different. Consider the example of a uniformly loaded cantilever (fig. 5.6.0).

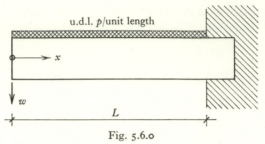

Fig. 5.6.0

Then
$$w = Ax^3 + Bx^2 + Cx + D + \frac{px^4}{4!\, \beta}. \tag{5.6.6}$$

This is as for an ordinary beam, but at $x = 0$, $F = M = 0$, from which

$$w''' = 0 \qquad \text{whence} \quad A = 0,$$

and $\qquad w'' = -(p/\mathcal{D}) \quad \text{whence} \quad 2B = -(p/\mathcal{D}).$ $\qquad (5.6.7)$

At the fixed end, $x = L$, we require $w = u = 0$ when

$$-\frac{p}{2\mathcal{D}} L^2 + CL + D + \frac{pL^4}{4!\,\beta} = 0$$

and $\qquad -\frac{p}{\mathcal{D}} L + C + \frac{\beta}{\mathcal{D}} \frac{pL}{\beta} + \frac{pL^3}{3!\,\beta} = 0.$ $\qquad (5.6.8)$

Hence
$$C = -\frac{pL^3}{3!\,\beta}$$

and $\qquad D = \frac{pL^4}{8\beta} \left(1 + \frac{4\beta}{\mathcal{D}L^2} \right).$ $\qquad (5.6.9)$

This latter result shows that the tip deflexion is increased from the pure bending value of $pL^4/8\beta$ by the value of the second term, $pL^2/2\mathcal{D}$.

As a proportional increase, the result can be expressed as follows. For a rectangular section $A = bd$, $\mathcal{D} = G(1 - \nu^2)A$, $I = Ad^2/12$ and if $\nu = 0.3$ then

$$\Delta\, \text{tip} = \text{tip deflexion}$$

$$= \frac{pL^4}{8\beta} \left(1 + \frac{4\beta}{\mathcal{D}L^2} \right)$$

$$= \frac{pL^4}{8\beta} \left(1 + 0.95 \left(\frac{d}{L} \right)^2 \right)$$

or $\qquad \Delta_{\text{second order}} \doteqdot \Delta_{\text{first order}} \times (1 + (d/L)^2).$ $\qquad (5.6.10)$

5.6 *Second order beam theory: shear deformation*

We can now see that the effects of shear deformation are likely to be small in most practical problems, where $d \ll L$; but if $d \doteqdot L$ the shear deformation effects are comparable to those arising from bending.

5.7 Approximate solution of differential equations by use of moments

In this section we shall discuss in an elementary way a procedure which can be adopted to obtain an approximate solution to a given problem described by a differential equation.

The example we shall consider is of a uniform column loaded by its own self weight. The requirement is to find the height consistent with stability. This problem was first considered by A. G. Greenhill in 1881 in connexion with the strength and stability of trees; the equation is of variable coefficient type and the solution can be expressed in terms of Bessel's functions, see Love (5.9). We shall however not use this information.

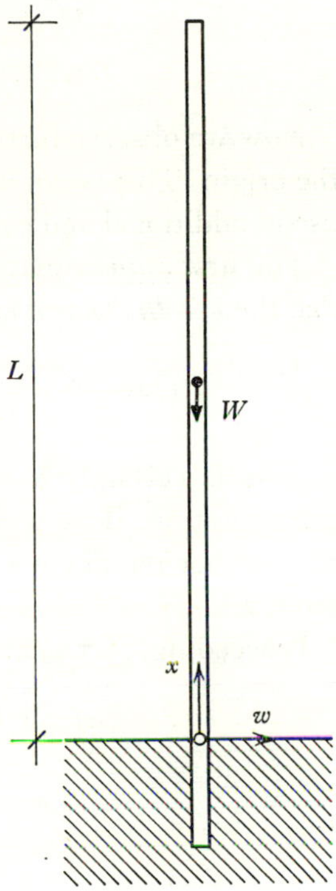

From (3.1.0–1) we have that

$$
\left.\begin{aligned}
M' + (W/L)(L-x)\,w' &= F, \\
F' &= p, \\
M &= \beta w''.
\end{aligned}\right\} \quad (5.7.0)
$$

Now $p = 0$ in this case, hence F is constant and since $F = 0$ at the free upper end, F is zero everywhere. Hence

$$ L(w) = \beta w''' + (W/L)(L-x)\,w' = 0 \quad (5.7.1) $$

is the governing equation. The associated boundary conditions are $w = w' = 0$, $x = 0$; $M = 0$, $x = L$. The plan is to attempt an approximate solution of (5.7.1) by use of moment concepts.

Fig. 5.7.0

The first decision in applying an approximation scheme is to choose a form for the dependent variable (here w) which contains a number of parameters. These parameters are usually taken to be constants. It is the purpose of the approximation scheme to find 'reasonable' values for the

12-2

parameters. The approximation scheme proceeds by reducing the differential equation to an algebraic equation.

In the present case we shall assume a polynomial form for w. As a preliminary, too, we shall attempt to choose the form for w so as to satisfy automatically as many boundary conditions as possible. Here the conditions at the origin are satisfied by an x^2 factor. Thus

$$w(x) = x^2(a + bx + cx^2 + dx^3) \qquad (5.7.2)$$

will be our assumed form, with four unknown constants, a–d. The remaining boundary conditions $M = 0$ $(F = 0)$ at $x = L$ imply that

$$
\begin{aligned}
w'' &= 2a + 6bL + 12cL^2 + 20dL^3 = 0, \\
w''' &= \qquad\quad 6b + 24cL + 60dL^2 = 0.
\end{aligned} \right\} \qquad (5.7.3)
$$

Now we observe further that if $b = 0$, the equation will be satisfied at the origin. This is not a necessary requirement but, as will be seen, is a useful additional and easily achieved condition.

The first approximation can now be found by applying the condition that the *zeroth moment of the entire equation* $(5.7.1)$ *should be zero*, namely

$$(L(w))^{(0)} = \frac{1}{L}\int_0^L (\beta w''' + (W/L)(L-x)w')\,dx = 0. \qquad (5.7.4)$$

This is certainly a necessary but not a sufficient condition for an exact solution. If we wish to improve the approximation we shall have to add further parameters in $(5.7.2)$ and require that further $(L(w))^{(i)} = 0$, $i = 1, 2, \dots$.

Proceeding, $(5.7.4)$ gives

$$
\left. \beta w'' \right|_0^L + \frac{W(L-x)}{L} \left. w \right|_L^0 - \int_0^L w\left(-\frac{W}{L}\right) dx = 0, \right\}
$$
$$
\text{or} \qquad\qquad\qquad -\beta w_0'' + \frac{W}{L}\int_0^L w\,dx = 0, \qquad (5.7.5)
$$

after the boundary conditions have been used.

Hence

$$-2a\beta + \frac{W}{L}\left(\frac{aL^3}{3} + \frac{cL^5}{5} + \frac{dL^6}{6}\right) = 0. \qquad (5.7.6)$$

Now $dL = -0.4c$, $cL^2 = -0.5a$ from $(5.7.3)$, whence in $(5.7.6)$

$$-2a\beta + WaL^2(\tfrac{1}{3} - \tfrac{1}{15}) = 0,$$

$$\text{or} \qquad\qquad L^2 = 7.5(\beta/W), \quad L = 2.74\sqrt{(\beta/W)}. \qquad (5.7.7)$$

The Bessel's function solution yields $L^2 = 7 \cdot 84(\beta/W)$ whence we see that this first approximation is only a few per cent in error. We have remarked that this approximate result can be improved by including further parameters and moment conditions.

Alternatively, if the *first* moment of (5.7.4) is set to zero instead of the zeroth, then the estimate for L is $2 \cdot 84\sqrt{(\beta/W)}$. We may then proceed by noting that the meaned estimate from the zeroth and first moments is likely to be more accurate than either singly. Hence we would obtain $2 \cdot 79\sqrt{(\beta/W)}$ to be compared with the accurate value of $2 \cdot 80\sqrt{(\beta/W)}$. This is clearly a very satisfactory outcome for a very small investment of effort. But not all results sought are as easily or satisfactorily evaluated.

5.8   Some further features of approximation procedures

The moment methods which are being advocated here take a global view of the approximation and the equation being approximated is probably not satisfied exactly anywhere in the domain.

An opposite extreme would be to satisfy the equation at isolated points only. Such a process is described as *collocation*. It is suggested that a satisfactory and efficient procedure is to combine these two extremes in most problems attempted, but to combine them via a least squares fit of an *over-determined system of algebraic equations*.

By an over-determined system of equations is meant a system of algebraic equations in which there are more equations than unknowns. Suppose there are n equations in m unknowns a_i and let these be written in matrix form as

$$\underset{(n \times m)}{A} \quad \underset{(m \times 1)}{\mathbf{a}} \quad = \quad \underset{(n \times 1)}{B} , \qquad (5.8.0)$$

where $m < n$. Equations (5.8.0) do not have a solution in general but it is shown in, for example, Turnbull and Aitken (6.8), that a least squares solution of this system in the sense that the squared inconsistency in the system (5.8.0) is minimized, requires that a solution vector $\mathbf{a}$ is computed from

$$A^T A \mathbf{a} = A^T B. \qquad (5.8.1)$$

The system (5.8.1) now has a unique solution since $A^T A$ is square. Also, $A^T A$ is *symmetric* – an important feature. However, not one of the original equations is satisfied exactly. By the methods we are proposing, A is in general unsymmetric when $m = n$.

As an alternative to this type of least squares fit we may attempt a least squares solution of the original equation. Thus we assume the dependent variable w to be expanded as

$$w(x) = \sum_i a_i \Phi_i(x) \quad (i = 1, ..., n), \tag{5.8.2}$$

where each $\Phi_i(x)$ is a known, selected function which satisfies all the boundary conditions and a_i are constants to be computed. If the equation to be solved approximately is denoted by

$$\xi(x) = 0, \tag{5.8.3}$$

then

$$\int (\xi(a_i \Phi_i(x)))^2 \, dx = \text{minimum} \tag{5.8.4}$$

results in the *least squares* conditions

$$\int \sum_{i=1}^n \xi(a_i \Phi_i(x)) \xi(\Phi_j(x)) \, dx = 0 \quad (j = 1, ..., n) \tag{5.8.5}$$

after differentiating (5.8.4) with respect to each a_i. This scheme will lead to a symmetric coefficient matrix $\mathbf{A}$ (5.8.0) when $m = n$. A generally simpler scheme is to adopt the *Galerkin* procedure and seek to find the a_i from

$$\int \sum_{i=1}^n \xi(a_i \Phi_i(x)) \Phi_j(x) \, dx = 0, \tag{5.8.6}$$

where again the coefficient matrix, $\mathbf{A}$, if square, is symmetric. Now it can be seen where the moment procedure fits into the general scheme since the moment requirement is merely

$$\int \sum_{i=1}^n \xi(a_i \Phi_i(x)) x^m \, dx = 0, \tag{5.8.7}$$

$m = 0, 1, ..., n-1$. Clearly here the coefficient matrix analogous to $\mathbf{A}$ in (5.8.0) will not be symmetric when square. This is a disadvantage. However, the motive for using (5.8.7) in preference to (5.8.5 or 6) is that in general there is a considerable saving in computational effort. The penalty incurred is that the results are neither quite as accurate nor as stable as those from (5.8.5 or 6) for given numbers of parameters. With skill, however, they are all comparable. But in dealing with *two dimensional* problems there is a definite advantage in the moment procedures in that suitable $\Phi_i(x)$ may be difficult if not impossible to construct. This circumstance hardly affects (5.8.7) although it means (5.8.5 and 6) cannot be used without modification.

5.8 *Further features of approximate procedures*

In conclusion we note that should we choose a Φ_i system which happens to contain, or be capable of describing, the exact solution to the problem in hand, then *any* of the methods mentioned (and others) will yield the exact solution. This is because the different approaches are all *necessary* conditions and the favourable choice of Φ_i will ensure that they are also *sufficient* conditions.

5.9 References

Any of the references asterisked would serve as a suitable introduction to Cartesian tensors. Pearson in his later chapters goes beyond our present needs.

Benjamin, J. R. *Statically Indeterminate Structures*, McGraw-Hill, New York (1959). Benjamin discusses problems where shear deformation is significant. But a full discussion based on the differential equations of the present chapter is not available outside the periodical literature. The theory is there sometimes referred to as 'Timoshenko beam theory'.

*Jeffreys, H. *Cartesian Tensors*, Cambridge University Press, London (1931).

Love, A. E. H. *The Mathematical Theory of Elasticity*, Cambridge University Press, London (1924). This reference is to the fourth edition of Love's great work. It remains the standard work on the topics discussed. The historical introduction and chapters on beam theory make an interesting study.

*Pearson, C. *Theoretical Elasticity*, Harvard University Press, Cambridge, Mass. (1959), chapters I and II.

*Temple, G. *Cartesian Tensors – An Introduction*, Methuen, London (1960).

Tiffen, R. *Plane Elastic Deformation*, Longmans Green, London (1970). Here general tensors, the elastic equations, and moments are all dealt with in an economical way, in chapters 1 and 5. The remainder of the book is beyond our present scope.

Washizu, K. *Variational Methods in Elasticity and Plasticity*, Pergamon, London (1968). Here the methods alternative to those used in our chapter 5 are presented, see especially chapter 7 of Washizu's book.

6

A MAXIMUM/MINIMUM THEOREM
FOR FRAMEWORKS

6.0   Introduction

We have been at some pains in the earlier chapters to emphasize the comparative age of many of the concepts used in our discussion of structural theory, while at the same time pointing out the novelty in the extent to which the concepts are used in the present treatment.

The present chapter contains some new results and the subject is not well represented in the literature. But even so there are links with the classical development of the theory. A difference in emphasis between this chapter and the earlier ones will also be evident in that the real use of the theorem we shall discuss is as a *design* tool whereas most of what has been said in previous chapters is primarily concerned with *analysis* of a given design.

The subject we shall discuss is a certain maximum/minimum property for structures and especially elastic structures. This is the first time since we outlined the limit theorems that such a property has been discussed. It is also interesting to observe that such properties tend to be associated in the literature with energy methods, which methods, for the reasons we have given in the introduction, we have avoided using. It can be concluded, therefore, that maxima and minima are not solely relevant to the realm of energy methods.

Although here our discussion of engineering structures has been at the level of beams and frameworks, it is anticipated that the reader will already have been exposed to a course in the theory of mechanics of continua; what is traditionally called 'strength of materials'. That is to say, it is expected that simple discussion of stress and strain in two dimensions, rather than as here of forces and displacements in one dimension, will have been encountered.

Central to such two dimensional discussion is the notion of principal value and direction, of stress and strain for example. The principal direction of stress is that direction through a given point across which the shear stress goes to zero. There are two such directions and they are *orthogonal.* The direct stress value then acting is either a maximum or a minimum of direct stress. The same is true of strain, and if the material is isotropic, then the principal directions of stress and strain coincide. If there is not complete isotropy then the principal stress directions will in general not coincide with the principal directions of strain, although each pair will be orthogonal. The discoveries of these concepts date back to the eighteenth and early nineteenth century.

The existence of principal stresses on orthogonal directions does not require that the stressed material be elastic; expressed in abstract terms the principal property is a consequence of the stress transforming for change of axes at a point as a symmetric second order tensor. The strain components also constitute a symmetric second order tensor, as do the components of second moment of area – the moments and products of inertia. The theorem we are about to state and prove is of a similar but somewhat different character since it is completely analogous to the stress and strain properties stated above only for elastic materials.

If the material is non-elastic then in general a distinction must be made between principal directions on the one hand and maximum/minimum directions on the other. This will be seen to be true for frameworks, and it is also true of the stress and strain behaviour of materials in which couple stresses or body couple are present. Such a situation has recently been discussed in the continuum mechanics literature.

6.1 The theorem: two-dimensional version

Consider a plane structure, a given but arbitrary point on it and an external vector test force, **F**, applied to the frame at this point. Let **u** be the displacement vector of this point of force application. In general the vectors **F** and **u** will not be collinear. The theorem states that, as the inclination of the force with respect to the structure is altered, there are just *two* directions of the force for which the two vectors are collinear. We shall call these directions *principal directions*. If the structure possesses sufficient symmetries, *all* directions through a point may be principal but we shall exclude this case.

We must distinguish between two types of behaviour possible, depending upon the structural response. Thus we shall suppose the displacement **u** to be related to the force **F** by a relation:

$$\mathbf{u} = \mathscr{A}\mathbf{F}, \tag{6.1.0}$$

in which $\mathscr{A}$ is a matrix (here 2×2) of *constant* elements. Such a relation is typical of elastic structures. For the vast majority of situations, too, $\mathscr{A}$ is a *symmetric* matrix. In stating the theorem we shall treat the cases of $\mathscr{A}$ symmetric and non-symmetric separately.

If $\mathscr{A}$ is a symmetric matrix then these two principal directions are *orthogonal* and the displacement components $(\mathbf{u}.\mathbf{F})/F$ along the force direction are either maximum or minimum. If $\mathscr{A}$ is not symmetric then the principal directions still exist but are *non-orthogonal* and the extremal values of $(\mathbf{u}.\mathbf{F})/F$ occur along yet other *orthogonal* directions.

The collinear displacement component, $(\mathbf{u}.\mathbf{F})/F$, is thus analogous to the direct stress in a two dimensional stressed material and the two directions for which

$$\delta = \left| \mathbf{u} - \frac{\mathbf{u}.\mathbf{F}}{F^2}\mathbf{F} \right|$$

goes to zero have been called 'principal directions' by analogy with the principal stress or strain directions.

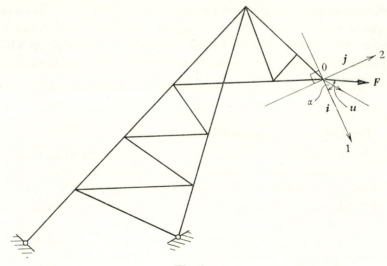

Fig. 6.2.0

6.2 Proof of the theorem

Various proofs are possible. We offer an elementary one. Consider the structure shown in fig. 6.2.0 and the point of force application, O. For some arbitrary choice of fixed orthogonal axes 1, 2 and associated unit vectors $\mathbf{i}$, $\mathbf{j}$ as indicated, the components of $\mathbf{F}$ along 1, 2 will be denoted by

$$F_1 = \mathbf{F}.\mathbf{i} = F\cos\alpha,$$
$$F_2 = \mathbf{F}.\mathbf{j} = F\sin\alpha, \qquad (6.2.0)$$

where
$$F = |\mathbf{F}|.$$

The displacement, $\mathbf{u}$, resulting from $\mathbf{F}$, will be resolved along the 1, 2 directions as

$$\mathbf{u} = \begin{bmatrix} u_1 \\ u_2 \end{bmatrix} = u_1\mathbf{i} + u_2\mathbf{j}$$

and then, in general,

$$\begin{bmatrix} u_1 \\ u_2 \end{bmatrix} = \begin{bmatrix} a_{11} & a_{12} \\ a_{21} & a_{22} \end{bmatrix} \begin{bmatrix} F_1 \\ F_2 \end{bmatrix}. \qquad (6.2.1)$$

We shall assume the a_{ij} to be material constants and independent of α. In the great majority of practical situations, $a_{21} = a_{12}$, namely the force/displacement relation is typified by a *symmetric* connexion matrix $\mathscr{A} = (a_{ij})$, which may be variously written as

$$u_i = (a_{ij}) F_j,$$
$$u = \mathscr{A}F, \quad \mathscr{A} = (a_{ij}). \qquad (6.2.2)$$

Now the component of displacement $\mathbf{u}$ along $\mathbf{F}$ will be denoted by Δ and is given by

$$\Delta = \frac{\mathbf{u}.\mathbf{F}}{F} = u_1\cos\alpha + u_2\sin\alpha$$

$$= F(a_{11}\cos^2\alpha + (a_{12}+a_{21})\sin\alpha\cos\alpha + a_{22}\sin^2\alpha). \qquad (6.2.3)$$

The displacement component at right angles to $\mathbf{F}$ will be denoted by δ and is given by

$$\delta = \left| \mathbf{u} - \frac{\mathbf{u}.\mathbf{F}}{F^2}\mathbf{F} \right| = u_2\cos\alpha - u_1\sin\alpha$$

$$= F((a_{22}-a_{11})\sin\alpha\cos\alpha + a_{21}\cos^2\alpha + a_{12}\sin^2\alpha). \qquad (6.2.4)$$

For the principal directions $\delta = 0$ and (6.2.4) then gives a quadratic for $t = \tan\alpha$, thus determining the *two* directions as the roots of

$$t^2 + \frac{a_{11}-a_{22}}{a_{12}}t - \frac{a_{21}}{a_{12}} = 0. \qquad (6.2.5)$$

It is clear that the directions are orthogonal only if $\mathscr{A}$ is symmetric, when $a_{12} = a_{21}$ and the product of the roots is then minus one.

The extremal property follows from (6.2.3) with $\partial\Delta/\partial\alpha = 0$ when, with $t = \tan\alpha$,

$$t^2 + 2\frac{a_{11} - a_{22}}{a_{12} + a_{21}}t - 1 = 0. \tag{6.2.6}$$

Hence the extremal property, the maximum or minimum for Δ, occurs on orthogonal directions even if $a_{12} \neq a_{21}$. In general these directions are different from either principal axis. If $a_{12} = a_{21}$, however, it is seen that the pairs of roots from (6.2.5, 6) coincide and hence, in this important class of physical problems, the principal directions are orthogonal and $\Delta = (\mathbf{u}.\mathbf{F})/F$ is extremal on these same axes.

The components $\Delta = (\mathbf{u}.\mathbf{F})/F$ and $\delta = |\mathbf{u} - (\mathbf{u}.\mathbf{F}/F^2)\mathbf{F}|$ are thus components of a second order tensor. In the case of symmetry, $a_{12} = a_{21}$, the Δ/δ plot is a circle, analogous to the Mohr circle of stress or strain. Such a plot is established by experimental observation in 6.4. If $a_{12} \neq a_{21}$ then the $\Delta|\frac{1}{2}(a_{12} + a_{21})$ plot is a circle from which it is evident that the extremal values of Δ are attained not when $\mathscr{A}$ has been transformed to a *diagonal* form but when transformed to a form such that the current $a_{12} = -a_{21}$.

Physically a_{11}, a_{22} must always be positive but a_{12}, a_{21} may take on either positive or negative values. Conceivably, if a_{12} and a_{21} differ in sign, then the principal axes may not exist.

6.3 An example

Consider the parabolic arch structure shown in fig. 6.3.0. We require to find the inclination of the principal axes at the cantilever tip O, and the extremal values of the displacement Δ. It should be remarked that principal axes as we have defined them exist at every point of a structure, with the orientation in general varying from point to point.

The arch will be assumed to have a secant variation of EI, namely $(EI)_{\text{at }s} = (EI)_{\text{at }o} \times \sec\psi$, where $(EI)_{\text{at }s}$ is used to denote EI evaluated at s. This is a feature of many practical arches. The defining equations for the linear elastic behaviour of the arch rib are, first, three equilibrium equations,

$$\left.\begin{array}{l} N_s + \kappa Q = 0, \\ Q_s - \kappa N = 0, \\ M_s - Q = 0, \end{array}\right\} \tag{6.3.0}$$

where $M_s = \mathrm{d}M/\mathrm{d}s$ and $\kappa = 1/R$ is the local curvature of the rib. The present problem is determinate and hence (6.3.0) are soluble independently of the displacements. We wish, however, to determine the dis-

6.3 An example

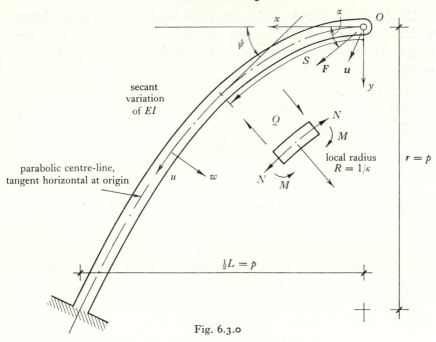

Fig. 6.3.0

placements and hence we require the material property equations. They are two in number, namely the bending moment/change of curvature relation

$$M = EI(w_s + \kappa u)_s, \qquad (6.3.1)$$

and the inextensional relation $u_s = \kappa w.$ $\qquad (6.3.2)$

The complete problem is thus defined by five equations (6.3.0, 1, 2) in five unknowns, N, Q, M; u, w. In terms of the rise and semi-span defined on fig. 6.3.0, the arch shape is described by

$$y = \frac{4r}{L^2}x^2, \quad y_x = \frac{8rx}{L^2}, \quad y_{xx} = \frac{8r}{L^2} = \text{const.} \equiv H = \frac{2}{p}. \quad (6.3.3)$$

Alternatively, the intrinsic coordinates s, ψ, the arc length and tangent angle respectively, are related to x, y by

$$\left.\begin{array}{c} \dfrac{\mathrm{d}\psi}{\mathrm{d}s} = \kappa, \quad \dfrac{\mathrm{d}y}{\mathrm{d}x} = \tan\psi, \quad \kappa = \dfrac{y_{xx}}{(1+(y_x)^2)^{\frac{3}{2}}}, \\[3mm] \kappa = \dfrac{y_{xx}}{\sec^3\psi} = H\cos^3\psi, \end{array}\right\} \qquad (6.3.4)$$

and $\qquad \left(\dfrac{1}{\kappa}\right)_{\psi} = \dfrac{\mathrm{d}(1/\kappa)}{\mathrm{d}\psi} = \dfrac{3\tan\psi}{\kappa}.$

189

Also $( )_s = \kappa( )_\psi$ but $( )_{s\psi} \ne ( )_{\psi s}$, that is the operations of differentiation with respect to s and ψ do not commute except when $\kappa = $ const. A subscript will be used as a shorthand for differentiation with respect to that subscript.

From $(6.3.0)_{1,2}$
$$N_{\psi\psi} + N = 0,$$

whence
$$N = A \sin\psi + B \cos\psi. \tag{6.3.5}$$

Again, from $(6.3.0)$

$$Q = -A \cos\psi + B \sin\psi$$

and
$$M = \frac{1}{H}\left(-A \tan\psi + \frac{B}{2}\sec^2\psi + C^*\right). \tag{6.3.6}$$

Instead of the third arbitrary constant C^*, $C = 1/H.(C^* + B/2)$ gives

$$M = \frac{1}{H}\left(-A \tan\psi + \frac{B}{2}\tan^2\psi\right) + C. \tag{6.3.7}$$

Then it is easily seen that A, B, C are respectively the shear, tension and bending moment at the origin of the coordinate system.

If w is eliminated from $(6.3.1, 2)$ and $(6.3.7)$ used, a second order equation for u is obtained which integrates to give

$$u = a \sin\psi + b \cos\psi + V_1 \sin\psi + V_2 \cos\psi, \tag{6.3.8}$$

where
$$V_1 = \int \cos\psi . r(\psi)\,d\psi, \quad V_2 = -\int \sin\psi . r(\psi)\,d\psi$$

and
$$r(\psi) = \frac{1}{\kappa H^2(EI)_0}\left(-\frac{A}{2}\sec^2\psi + \frac{B\tan^3\psi}{6} + CH\tan\psi + D\right),$$

$(EI)_0$ being the section stiffness at the coordinate origin (fig. 6.3.0). The tangential displacement u is thus expressible in terms of six arbitrary constants a, b; A, B, C, D.

Finally w, the transverse displacement, is given by

$$w = u_s/\kappa = u_\psi = a \cos\psi - b \sin\psi + V_1 \sin\psi - V_2 \cos\psi. \tag{6.3.9}$$

The conditions at the fixed end are

$$u = u_\psi = u_{\psi\psi} = 0, \quad \text{at} \quad y = p. \tag{6.3.10}$$

These conditions require that

$$\begin{aligned}
H^3(EI)_0 b + (\tfrac{25}{8}A - \tfrac{16}{15}B - \tfrac{5}{2}D) &= 0, \\
H^3(EI)_0 a - (\tfrac{7}{3}A - \tfrac{2}{3}B - 2D) &= 0, \\
2{\cdot}5A - \tfrac{4}{3}B - D &= 0.
\end{aligned} \tag{6.3.11}$$

At the free end, O, $M = 0$ from which $C = 0$.

The displacement/force relation at the free end can now be formulated as

$$\mathbf{u}_0 = \begin{bmatrix} u_0 \\ w_0 \end{bmatrix} = [\mathscr{A}] \begin{bmatrix} -B \\ -A \end{bmatrix} = \mathscr{A}.\mathbf{F}, \tag{6.3.12}$$

where

$$u_0 = b + (V_2)_0 = b - \frac{1}{H^3(EI)_0} \left(-\frac{A}{8} + \frac{D}{2} \right),$$

$$w_0 = a \quad \text{and} \quad \mathbf{F} = -\begin{bmatrix} B \\ A \end{bmatrix}.$$

Using (6.3.11) in (6.3.12) we obtain

$$\mathbf{u} = \begin{bmatrix} u_1 \\ u_2 \end{bmatrix}_{\text{at } 0} = \begin{bmatrix} u \\ u_y \end{bmatrix}_{\text{at } 0} = \frac{1}{H^3(EI)_0} \begin{bmatrix} \frac{8}{5} & -2 \\ -2 & \frac{8}{3} \end{bmatrix} \begin{bmatrix} -B \\ -A \end{bmatrix}$$

$$= \mathscr{A}.\mathbf{F}, \tag{6.3.13}$$

from which we see that $\mathscr{A}$ is indeed symmetric. The principal directions are the roots of the quadratic (6.2.5 or 6) with $a_{12} = a_{21}$. Namely

$$t^2 + \tfrac{8}{15}t - 1 = 0, \tag{6.3.14}$$

or

$$t = 0.768, \quad -1.302,$$

$$\alpha_1 = 37° 30', \quad \alpha_2 = \tfrac{1}{2}\pi + \alpha_1. \tag{6.3.15}$$

The corresponding values for the extremals of the displacement Δ are

$$\Delta = \frac{F(8 - (6/t))}{3H^2(EI)_0} = (0.065, 4.21)\frac{F}{H^3(EI)_0}. \tag{6.3.16}$$

This example is not of any particular intrinsic interest except that it is considered by Matheson and Francis (4.10). Their solution reads '37° 30', clockwise from the vertical'. It is probably safe to conclude that they were unaware of the existence of a second orthogonal direction which also exhibits a principal property, the property they set out to study. There is also no indication regarding the extremal properties of Δ.

Actually for their solution to be correct either the sense of rotation they indicate should be reversed, then giving the weak axis, or else the angle quoted should be complemented with $\tfrac{1}{2}\pi$, i.e. $\tfrac{1}{2}\pi - 37° 30'$, and then used with the clockwise rotation, giving the strong direction.

Exercises

(1) We leave it as an exercise to show that the Δ/δ plot for variable α is a circle as indicated in fig. 6.3.1. The convention used in drawing this circle is that when looking along the direction of $\mathbf{F}$, δ is positive to the right. This convention preserves the direction of rotation from the physical to the Mohr plane.

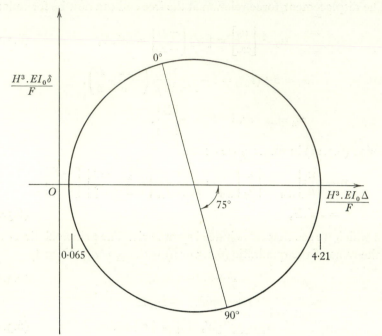

$\dfrac{H^3 . EI_0 \delta}{F}$

O

$\dfrac{H^3 . EI_0 \Delta}{F}$

$75°$

$0°$

$90°$

0.065

4.21

Fig. 6.3.1

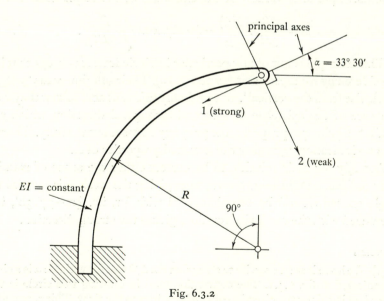

principal axes

$\alpha = 33° 30'$

1 (strong)

2 (weak)

EI = constant

R

$90°$

Fig. 6.3.2

192

(2) For the quadrant circular arch member shown in fig. 6.3.2, show that the principal axis inclinations are given by the roots of $t^2 + (4 - \pi) t - 1 = 0$ where $t = \tan \alpha$, and that the extremal displacements Δ_1, Δ_2 are

$$\Delta_1 = 0 \cdot 021, \quad \Delta_2 = 1 \cdot 108 \quad \left(\times \frac{FR^3}{EI} \right).$$

Various points emerge from an examination of the results. First the ratio of the extremal displacements. In the case of the parabolic rib this ratio is nearly 1 to 65. Secondly, we note that the strong axis intersects the member so as to ensure that part of the member is bent one way and part the other, hence minimizing the tip displacement. The weak axis on the other hand is oriented so as to bend the member as much like a cantilever as possible.

This rather high ratio for extreme values of Δ means that the member is to a first approximation rigid in extension along the Δ_1 direction and surprisingly flexible along the Δ_2 direction. In the following section we give some experimental results for a perspex model of the parabolic member studied in 6.3.

The external forces and moments which deform a plane frame in its plane are first, forces in the plane such as we have been considering; secondly *moments* about axes *normal* to the plane of the frame. The tip of the cantilever, O, in 6.3 rotates when the force **F** is applied along a principal direction. Thus the principal directions are principal only for *forces* and *displacements*, but, with this restriction, every point in every structure has a pair of principal axes passing through it. If we seek to generalize the 'principal' concept to include moments and rotations then we in general can only achieve this by imagining a stiff weightless lamina attached to the member at the point concerned and displacing and rotating with it. Then in general there is *one* point in this lamina and not coinciding with the point of attachment, which exhibits 'principal' behaviour with respect to forces acting in a specified pair of orthogonal directions and a moment applied about an axis normal to the lamina at this point. Such a point is commonly termed an *elastic centre*. This point will change as the inclination of the specified axes is changed.

6.4   An experimental result

The parabolic arch discussed in 6.3 has been studied on a model scale and we present now a summary of the findings. The model was constructed from 4·6 mm thick perspex which was found to have a short term Young's modulus of $31 \times 10^8 \, N/m^2$ as measured using a cantilever beam cut from the same sheet. In the notation of fig. 6.3.0, in the model, $r = L/2 = p = 0.20$ m and the section depth at O was 0.0105 m. With a secant variation of EI this required a section depth at the root of the curved cantilever of 0.0138 m. The test force $\mathbf{F}$ applied at O was $4.18 \, N$.

The displacement of the loaded point was followed by focusing a fine beam of light through the hollow load pin at O and marking the initial and displaced positions of O.

Some experimental values are as follows. A positive δ value indicates that the point O, when viewed from the line of action of $\mathbf{F}$ and looking toward O, moved to the left. The angles are measured from the negative x axis as $0°$ and anticlockwise positive.

Table 6.4.0

Angle (°)	Δ (mm)	δ (mm)
o	5·0	6·3
40	0·0	0·0
90	8·2	−6·1
−40	12·4	3·0
−80	10·8	−5·0

Now

$$I_0 = \frac{0.46 \times (1.05)^3}{12}$$

$$= 0.0443 \, \text{cm}^4.$$

Hence

$$\Delta_2 = \frac{4.21}{8} \times \frac{20^3 \times 4.18}{31 \times 10^4 \times 0.0443}$$

$$= 1.29 \, \text{cm},$$

$$\Delta_1 = \frac{1.29}{65} = 0.02 \, \text{cm}.$$

The theoretical circle and experimental points are shown plotted on fig. 6.4.0. The plot, clearly, must always lie to the right of the axis, since the force must always produce a Δ displacement in the same direction as the force.

It is seen that the agreement is very satisfactory considering the very simple equipment of model, scale, protractor, light source, spring loading device, pencil and sheet of paper. The weak principal axis was observed between $-55°$ and $-50°$ and the strong axis at between $35°$ and $40°$, both of which values span the theoretical values.

6.4 *An experimental result*

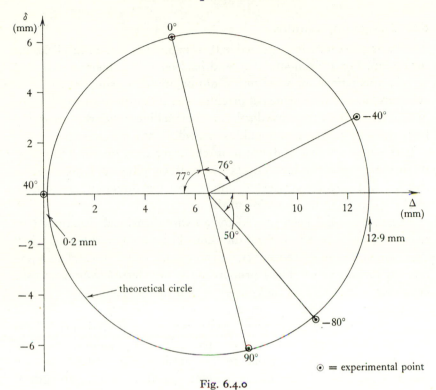

Fig. 6.4.0

Typical polar plots for Δ and δ are as indicated in fig. 6.4.1. This experimental method is presumably one of the simplest demonstrations of a second order tensor quantity plotting as a (Mohr) circle.

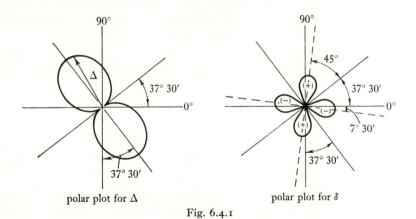

polar plot for Δ polar plot for δ

Fig. 6.4.1

6.5   Design implications

The theorem we have investigated in this chapter has various implications for design. For a given structure and load point the theorem says that certain orientations mobilize maximum or minimum support from the structure, namely the principal orientations. Again in mechanical component design if the response displacement is collinear with the disturbing force then the restraint from guides etc. will be at a minimum.

If the applied force is dynamic in character then motion transverse to the force may produce effects of mode coupling and possible instabilities. If the force is oriented along a principal direction however the total response is in the direction of the force.

The theorem has been proved for a two dimensional situation. It is however equally true in three dimensions but now with three principal axes and with three extrema of Δ, a maximum, a minimum and an intermediate turning value. These properties can be inferred from the tensor character of $(\mathbf{u}.\mathbf{F})/F$ and the components at right angles.

Consider a straight cantilever member of arbitrary cross-sectional shape. If the usual strength of materials assumption of section rigidity is invoked, then the position of an axis parallel to the cantilever axis such that forces acting perpendicular to the cantilever axis and intersecting this axis produce no twist of the cantilever, is discussed in strength of materials texts under the title of shear centre or twist centre. The cantilever loaded by a force in this way will bend but not twist – the elastic components of displacement due to bending being compounded of displacements parallel to the respective *principal inertia axes*. In terms of the type of principal axis discussed in this chapter, the *shear centre* position is characterized by having the $\mathbf{u}/\mathbf{F}$ principal axes parallel to the inertia principal axes of the (assumed) rigid cross-section.

If the usual assumption of rigidity of section shape is not made then an analogous point can still be found but we leave for discussion elsewhere the features of this problem.

In the present discussion we have taken an existing structure and found the principal axes. In a true design situation the knowledge afforded by the theorem would be used to influence the actual choice of structure. How this could be achieved would depend upon the type of design required but the probable way would be to use the knowledge to modify a design, perhaps to alter the inclination of the principal axes so as to bring them closer into line with the load direction. A simple design situation is depicted in fig. 6.5.0. It is required to provide a supporting structure to span AB but avoid the obstacle C. A possible design is sketched in. If we view the half structure as a cantilever it is seen that the strong axis is in the general direction indicated. Suppose we were to

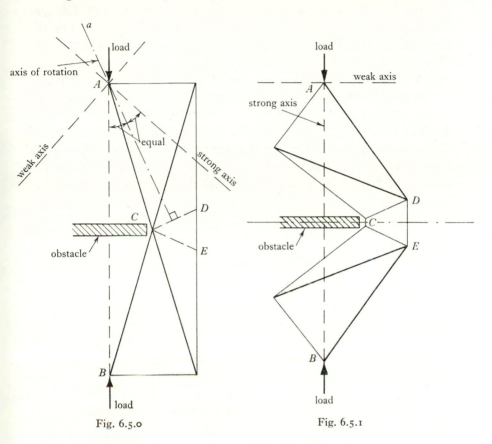

Fig. 6.5.0   Fig. 6.5.1

introduce additional members CD, CE; then consider CDA as a cantilever structure rooted at CD. The foregoing discussion would suggest that if the structure CDA was to be rotated about the axis aA indicated into the new position shown in fig. 6.5.1, and similarly for CEB, then a stiffer structure would be produced since the load axis would then coincide with the strong principal axis. An assumption here is that instability is not imminent in any of the members of the frame either as originally proposed or modified and that the portion CDE forms a rigid substructure, compared with the remainder.

The modified structure might be less desirable from strength considerations however since the forces in three of the members have changed sign and where before there were two compression and two tension members there are three compression and one tension member in the modified structure. In addition, the extra members *CD*, *CE* are required in the modified design. On the other hand, with the line of action of the load fixed, the original structure is such that the portion *CDE* displaces to the right under load. In the case of the structure as modified (fig. 6.5.1) there is no tendency for *CDE* to displace. Such a feature may be important for certain types of machine components and can be systematically achieved by exploiting the principal axis concept. Thus the theorem relates to stiffness and not to strength of the structure.

In many situations in practice the principal axes lie along the axis of a member. This is because of the near (but assumed) inextensibility of members according to the usual framework theory, such as discussed in chapters 1 and 2.

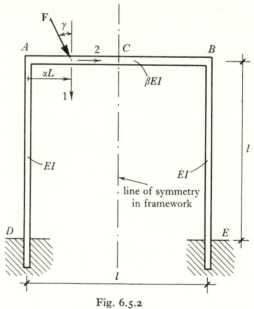

Fig. 6.5.2

Thus in the portal frame (fig. 6.5.2) the principal axes at points *A* and *B* are: strong axis vertical, weak axis horizontal, because of the column inextensibility. At *C* the axes are again horizontal and vertical, although either axis could be the strong axis depending upon the relative stiffnesses

198

of beam to column (β). But the vertical axis is more likely to be the strong axis. The reason now for the principal directions being horizontal and vertical is that C is a point of symmetry in the framework. For points on the columns other than A, D; B, E the axes remain vertical (strong) and horizontal (weak) because of column inextensibility.

For points on the beam however other than A, C, and B the axes are inclined. If $\beta = 1$ and with the axes 1, 2 as indicated in fig. 6.5.2, it can be shown that

$$\mathbf{u} = \begin{bmatrix} u_1 \\ u_2 \end{bmatrix} = \begin{bmatrix} a_{11} & a_{12} \\ a_{21} & a_{22} \end{bmatrix} \begin{bmatrix} F_1 \\ F_2 \end{bmatrix}, \qquad (6.5.0)$$

where
$$a_{11} = \frac{l^3}{84\beta} \alpha^2 (1-\alpha)^2 (13 + 4\alpha - 4\alpha^2),$$

$$a_{12} = a_{21} = \frac{l^3}{28\beta} \alpha(1-\alpha)(1-2\alpha),$$

$$a_{22} = \frac{5l^3}{84\beta}.$$

If $\alpha = \frac{1}{4}$ then $\gamma = -3\cdot6°$ and $\frac{1}{2}\pi - 3\cdot6°$ from which we see that the inclination is small. The $-3\cdot6°$ axis will be the strong principal axis. The nearness of the inextensible columns has a strong influence on the amount by which the axis inclination can depart from the vertical.

As yet we have not discussed an example in which the $a_{ij} \neq a_{ji}$. In fact such problems are difficult to find. All the describing equations which we have discussed are of the self-adjoint type (see Ince (0.8)) and this in turn implies symmetry of the a_{ij}'s. An exceptional circumstance is a structure on the point of collapse by hinge formation. We shall not pursue this topic here except to note that when a bar of unsymmetrical section is bent and twisted as a cantilever, as the collapse load is reached then principal axes of bending exist such that the load $\mathbf{F}$ producing the collapse and the resulting displacement $\mathbf{u}$ are collinear and it is the case that these axes are *not orthogonal* in general (Brown (6.8)). The implication is that $\mathscr{A}$ is unsymmetric. In this case however we are unable to write down a $\mathbf{u} = \mathscr{A}\mathbf{F}$ relation and the second part of the theorem which predicts orthogonal axes for $(\Delta_1, \Delta_2)_{\text{max. min.}}$, cannot be verified.

A genuine example of a structure with a $\mathbf{u} = \mathscr{A}\mathbf{F}$ relation with $\mathscr{A}$ unsymmetric probably requires a dissipative material. The situation is the counterpart to the so-called 'non-conservative' stability problems which have recently been discussed in the literature (see Ziegler (6.8)).

6.6   An alternative treatment of the theorem: matrices

The theorem we have discussed in this chapter can be looked at from a different point of view. This is the point of view which emphasizes the algebra of the problem and works in terms of matrices and, ultimately, of tensors. Thus far these notions have only been alluded to. Here we shall give a brief résumé of the properties from this point of view but some knowledge of matrix algebra and elementary tensor notions is assumed. If the reader wishes he may read the background material in Jeffries or Temple (5.9).

The crucial equation from which all the properties of principal behaviour and maxima/minima follow is (6.2.2), and especially the matrix of coefficients $\mathscr{A} = [a_{ij}]$. The next most important feature is the symmetry of the matrix $\mathscr{A}$, namely $a_{ij} = a_{ji}$. In the respects that we are dealing with a matrix a_{ij}, the principal properties are to be expected. For, given a symmetric (square) $n \times n$ matrix of real elements, this matrix can be reduced to diagonal, or so-called canonical form, by the following operations:

$$H\mathscr{A}H^T = H\mathscr{A}H^{-1} = D = \text{diagonal matrix.} \qquad (6.6.0)$$

Here H is an *orthogonal* $n \times n$ matrix namely $H^{-1} = H^T$, H^T is the transpose and H^{-1} the inverse of H. A unique H exists which will achieve this. The resulting diagonal matrix has a series of numbers down the diagonal which are called the eigenvalues, and these numbers are *real* by virtue of the *symmetry* of $\mathscr{A}$.

The meaning of (6.6.0) is that for the problem discussed in 6.2, if the orthogonal reference axes are changed to a new orthogonal set, then there exists a unique set for which the relation between force and displacement becomes of diagonal form – the *principal directions* of our earlier study. Our particular physical interest is in cases $n = 2$, 3, namely plane and spatial structures. The numbers on the diagonal of the matrix D (and of $\mathscr{A}$) are such that they are always positive. In addition the greatest and least of the numbers in D are such that no values which occur on the leading diagonal of any $\mathscr{A}$ can be greater than or less than these two values. All the matrices $\mathscr{A}$ are in fact a special class of matrices, they are *positive definite matrices*. A matrix is positive definite when leading minors of all orders are positive.

The form (6.6.0) arises because if we change axes, and if the forces transform as

$$\bar{\mathbf{F}} = H.\mathbf{F} \qquad (6.6.1)$$

then the displacements transform as

$$\mathbf{u} = H^T \bar{\mathbf{u}}. \tag{6.6.2}$$

Any $\mathbf{u}$, $\mathbf{F}$ which transform in this way to $\bar{\mathbf{u}}$, $\bar{\mathbf{F}}$ are spoken of as contragredient variables. In general coordinates if the F_j are components of a covariant vector then the u^i are components of a contravariant vector.

Then from $(6.6.1, 2)$ and if H is an orthogonal matrix

$$H^T.\bar{\mathbf{u}} = \mathbf{u} = \mathscr{A}.\mathbf{F}$$

or

$$\bar{\mathbf{u}} = H\mathscr{A}\mathbf{F} = H\mathscr{A}H^{-1}.\bar{\mathbf{F}}$$

$$= (H^{-1}\mathscr{A}H)^T.\bar{\mathbf{F}} \tag{6.6.3}$$

if $\mathscr{A}$ is *symmetric*.

If in addition the H is such as to reduce $H^{-1}\mathscr{A}H$ to diagonal form then

$$(H^{-1}\mathscr{A}H)^T = H\mathscr{A}H^{-1} = H^{-1}\mathscr{A}H. \tag{6.6.4}$$

Such transformations are important special cases of more general similarity transformations.

One of the earliest accounts which draws attention to the contragredient behaviour of $(\mathbf{u}, \mathbf{F})$ is Jenkins, (6.8). The product of $\mathbf{u}^T.\mathbf{F}$ is an invariant to such rotations and defines a work function.

There are other situations in which principal behaviour occurs. For symmetric connexion matrices analogous to $\mathscr{A}$ there are two subclasses of such phenomena. The first class is that with $\mathscr{A}$ positive definite. This class includes the present $\mathbf{u} = \mathscr{A}\mathbf{F}$ relations, together with the inertia matrix $I = [I_{xy}]$, the metric matrix for length measurement and the matrix of coefficients in quadratic functions associated with energy conservation.

The second subclass, which includes the first class as a special case, comprises the matrices associated with the usual stress, strain and curvature, among others. These are *symmetric* matrices but not positive definite. The differences between the two subclasses are best demonstrated on the Mohr plane in two dimensions when we plot a_{ii}/a_{ij}. The first class always plot in the *right half* of the Mohr plane whereas the larger second class can plot to any circle in the Mohr plane. In either case of course the circle centre will be on the a_{ii} axis.

The class of phenomena associated with unsymmetric matrices is potentially much wider than that for symmetric systems, but this type of matrix is encountered very much less frequently. The same is true of cases where the elements of $\mathscr{A}$ are not constants. Such problems are associated, for example, with couple stresses, body couple, magnetic effects and large displacements in some circumstances. The mention in 6.2 that principal axes may not exist is a manifestation of possible complex eigenvalues for the asymmetric $\mathcal{A}$ matrix.

Further development of our subject soon leads us to consider continua in two and three dimensions, where here we have been concerned with one dimension. Hence two and three dimensional coordinate systems will be needed. Also quantities analogous to $\mathcal{A}$, but with more indices implied, will be encountered. If we restrict attention to rectangular Cartesian coordinate systems then matrices and Cartesian tensors are the appropriate tools. If however more general, perhaps non-orthogonal, systems are encountered, then general curvilinear tensors are appropriate. We shall not pursue this topic here but plan to elsewhere.

There are comparatively few discussions in the theory of structures literature of any of the topics raised in this chapter. Some of the topics are, however, discussed in a recent book by Gregory, (6.8).

6.7   Conclusion

The theorem discussed in this chapter surprisingly seems not to have been discussed in the literature. It represents one of a class of results which can be constructed from vector valued quantities, here the force and displacement at a point. The components of the one vector along the direction of and normal to the other vector are shown to possess second order tensor properties and, as a result, all the properties here discussed follow. Clearly other associations of vectors are possible and may prove to be useful. Again, vector quantities at two *different* points might be considered. That the present theorem yields information about *structure stiffness*, too, provides some unity with the earlier parts of the book, in which an attempt has been made to raise the status of *stiffness* concepts on to a level nearer that usually accorded to *strength* concepts.

6.8   References

Brown, E. H. *Structural Analysis*, vol. 1, Longmans Green, London (1967). The main emphasis is on virtual work methods. The particular interest in connexion with our chapter 6 is the discussion on p. 196 and subsequently of plastic bending of beams with no axis of symmetry.

Gregory, M. S. *Introduction to Extremum Principles*, Butterworth, London (1969). The discussion of extremum principles in this book is much wider

6.8 *References*

than ours. Some of the features of our particular problem are discussed by Gregory on p. 164.

Jenkins, R. S. *Theory and Design of Cylindrical Shell Structures*, The O.N.Arup Group of Consulting Engineers, London (1947). This pioneering and difficult book discusses co- and contra-gredience in the chapter on 'The Statically Indeterminate Structure'. Much of the subsequent development of flexibility methods stems from this work.

Turnbull, H. W. and Aitken, A. C. *An Introduction to the Theory of Canonical Matrices*, Blackie, Glasgow (1932). A standard work. Chapter 4 is that most relevant to our discussion.

Ziegler, H. *Adv. Appl. Mechanics* (1956), **4**, 351.

APPENDIX OF TABLES

Table 2.25.0 *Results for shear wall structures*

n	P	Q	R	S	W_1	W_2
		$\mu = 0.01250$				
5	0.066	1.193	−4.743	8.17	1.12	22.58
7	0.096	1.370	−6.428	13.75	0.95	10.41
9	0.132	1.616	−7.944	19.62	0.81	5.98
11	0.174	1.943	−9.276	25.30	0.69	3.98
13	0.226	2.367	−10.417	30.55	0.60	2.94
15	0.288	2.910	−11.374	35.28	0.52	2.34
17	0.365	3.599	−12.156	39.48	0.45	1.97
19	0.460	4.468	−12.779	43.20	0.39	1.73
21	0.578	5.562	−13.261	46.49	0.35	1.56
23	0.725	6.934	−13.620	49.42	0.31	1.44
25	0.909	8.655	−13.873	52.03	0.27	1.35
27	1.138	10.809	−14.038	54.38	0.24	1.29
29	1.424	13.506	−14.128	56.50	0.22	1.24
31	1.781	16.880	−14.157	58.43	0.20	1.20
33	2.228	21.100	−14.135	60.19	0.18	1.17
		$\mu = 0.02500$				
5	0.138	1.397	−4.498	14.48	1.00	10.59
7	0.212	1.782	−5.907	22.73	0.79	5.05
9	0.308	2.346	−7.038	30.31	0.63	3.11
11	0.434	3.146	−7.898	36.87	0.51	2.26
13	0.604	4.263	−8.509	42.41	0.42	1.81
15	0.835	5.809	−8.909	47.07	0.35	1.56
17	1.150	7.939	−9.134	51.02	0.29	1.40
19	1.581	10.868	−9.220	54.41	0.25	1.30
21	2.170	14.891	−9.199	57.34	0.21	1.23
23	2.978	20.412	−9.094	59.92	0.19	1.18
25	4.086	27.987	−8.928	62.20	0.16	1.14
27	5.605	38.378	−8.715	64.23	0.14	1.11
29	7.688	52.631	−8.468	66.06	0.13	1.09
31	10.545	72.180	−8.196	67.72	0.12	1.08
33	14.463	98.991	−7.904	69.22	0.10	1.07
		$\mu = 0.05000$				
5	0.303	1.841	−4.037	23.68	0.82	5.15
7	0.507	2.742	−4.985	33.93	0.59	2.73
9	0.813	4.199	−5.546	42.11	0.44	1.90
11	1.284	6.506	−5.789	48.53	0.34	1.53
13	2.015	10.130	−5.794	53.65	0.27	1.34
15	3.154	15.805	−5.630	57.81	0.22	1.23
17	4.932	24.681	−5.350	61.28	0.18	1.16
19	7.708	38.555	−4.993	64.22	0.15	1.12
21	12.046	60.237	−4.585	66.74	0.13	1.09
23	18.822	94.116	−4.144	68.95	0.11	1.07
25	29.410	147.053	−3.682	70.88	0.09	1.06
27	45.953	229.769	−3.207	72.60	0.08	1.05
29	71.802	359.012	−2.722	74.13	0.07	1.04
31	112.191	560.955	−2.232	75.50	0.06	1.04
33	175.298	876.491	−1.739	76.73	0.06	1.03

Table 2.25.0 (Cont.)

n	P	Q	R	S	W₁	W₂
		$\mu = 0.10000$				
5	0.722	2.879	−3.212	34.91	0.62	2.79
7	1.399	5.287	−3.485	45.61	0.40	1.74
9	2.649	9.863	−3.341	53.28	0.28	1.38
11	4.985	18.483	−2.936	59.01	0.21	1.22
13	9.365	34.681	−2.379	63.47	0.16	1.14
15	17.584	65.098	−1.737	67.06	0.13	1.10
17	33.013	122.205	−1.050	70.03	0.10	1.07
19	61.978	229.416	−0.338	72.51	0.08	1.05
21	116.353	430.688	0.387	74.63	0.07	1.04
23	218.432	808.541	1.120	76.45	0.06	1.04
25	410.069	1517.896	1.856	78.03	0.05	1.03
27	769.834	2849.586	2.594	79.42	0.05	1.03
29	1445.229	5349.603	3.333	80.64	0.04	1.02
31	2713.166	10042.949	4.072	81.73	0.04	1.02
33	5093.507	18853.925	4.812	82.71	0.03	1.02
		$\mu = 0.20000$				
5	1.981	5.639	−1.837	46.34	0.42	1.77
7	4.858	13.606	−1.269	56.22	0.26	1.31
9	11.817	33.002	−0.389	62.98	0.17	1.16
11	28.701	80.120	0.628	67.94	0.12	1.10
13	69.694	194.538	1.703	71.75	0.09	1.06
15	169.229	472.369	2.803	74.77	0.07	1.05
17	410.917	1146.988	3.913	77.22	0.06	1.04
19	997.774	2785.074	5.026	79.25	0.05	1.03
21	2422.757	6762.610	6.142	80.95	0.04	1.03
23	5882.857	16420.746	7.258	82.39	0.03	1.02
25	14284.560	39872.304	8.374	83.64	0.03	1.02
27	34685.226	96816.406	9.490	84.72	0.02	1.02
29	84221.500	235086.344	10.607	85.67	0.02	1.02
31	204503.437	570827.751	11.723	86.51	0.02	1.02
33	496568.375	1386064.753	12.840	87.25	0.02	1.02
		$\mu = 0.40000$				
5	6.759	14.633	0.296	56.52	0.27	1.34
7	23.510	50.755	1.812	65.23	0.15	1.14
9	81.638	176.204	3.478	71.07	0.10	1.07
11	283.449	611.773	5.187	75.27	0.07	1.05
13	984.130	2124.061	6.909	78.42	0.05	1.04
15	3416.878	7374.689	8.634	80.87	0.04	1.03
17	11863.304	25604.707	10.360	82.83	0.03	1.03
19	41189.132	88898.984	12.087	84.43	0.03	1.02
21	143007.719	308655.250	13.813	85.76	0.02	1.02
23	496519.563	1071644.003	15.540	86.88	0.02	1.02
25	1723904.752	3720723.505	17.267	87.83	0.02	1.02
27	5985346.013	12918242.027	18.993	88.66	0.01	1.02
29	20780988.054	44851848.109	20.720	89.39	0.01	1.02
31	72151120.187	155724608.437	22.446	90.02	0.01	1.02
33	250507040.437	540672257.499	24.173	90.59	0.01	1.01

Table 2.25.1 *Results for shear wall structures (μ order of unity)*

n	P	Q	R	S	W_1	W_2
		$\mu = 1{\cdot}00000$				
1	1·000	2·000	0·000	25·00	1·67	10·67
2	3·000	5·000	0·667	43·33	0·63	2·50
3	8·000	13·000	1·875	54·49	0·33	1·48
4	21·000	34·000	3·333	61·91	0·20	1·21
5	55·000	89·000	4·891	67·27	0·14	1·11
6	144·000	233·000	6·486	71·33	0·10	1·08
		$\mu = 2{\cdot}00000$				
1	2·000	3·000	1·000	33·33	1·33	6·67
2	8·000	11·000	3·000	51·52	0·43	1·75
3	30·000	41·000	5·533	61·79	0·21	1·21
4	112·000	153·000	8·214	68·50	0·13	1·09
5	417·999	570·999	10·933	73·24	0·08	1·06
6	1559·996	2130·995	13·662	76·76	0·06	1·04
		$\mu = 3{\cdot}00000$				
1	3·000	4·000	2·000	37·50	1·17	5·33
2	15·000	19·000	5·200	55·26	0·35	1·50
3	72·000	91·000	8·875	65·11	0·17	1·13
4	344·999	435·999	12·643	71·46	0·10	1·06
5	1652·995	2088·993	16·430	75·88	0·06	1·04
6	7919·965	10008·957	20·220	79·14	0·04	1·04
		$\mu = 4{\cdot}00000$				
1	4·000	5·000	3·000	40·00	1·07	4·67
2	24·000	29·000	7·333	57·47	0·30	1·37
3	140·000	169·000	12·086	67·06	0·14	1·09
4	815·998	984·997	16·902	73·18	0·08	1·05
5	4755·982	5740·978	21·728	77·41	0·05	1·04
6	27719·890	33460·867	26·556	80·49	0·04	1·03
		$\mu = 5{\cdot}00000$				
1	5·000	6·000	4·000	41·67	1·00	4·27
2	35·000	41·000	9·429	58·94	0·27	1·30
3	240·000	280·999	15·229	68·36	0·12	1·07
4	1644·996	1925·995	21·076	74·31	0·07	1·04
5	11274·962	13200·957	26·929	78·40	0·05	1·03
6	77279·703	90480·640	32·783	81·38	0·03	1·03
		$\mu = 6{\cdot}00000$				
1	6·000	7·000	5·000	42·86	0·95	4·00
2	48·000	55·000	11·500	60·00	0·25	1·25
3	377·999	432·999	18·333	69·28	0·11	1·05
4	2975·993	3408·992	25·202	75·12	0·06	1·03
5	23429·933	26838·925	32·074	79·11	0·04	1·03
6	184463·406	211302·312	38·947	82·00	0·03	1·03

Table 2.25.1   (*Cont.*)

n	P	Q	R	S	W1	W2
		$\mu = 7\cdot00000$				
1	7·000	8·000	6·000	43·75	0·92	3·81
2	63·000	71·000	13·556	60·80	0·23	1·21
3	559·999	630·999	21·413	69·98	0·10	1·04
4	4976·989	5607·988	29·297	75·72	0·06	1·03
5	44232·867	49840·843	37·184	79·64	0·04	1·03
6	393118·438	442959·188	45·071	82·47	0·03	1·03
		$\mu = 8\cdot00000$				
1	8·000	9·000	7·000	44·44	0·89	3·67
2	80·000	89·000	15·600	61·42	0·22	1·19
3	791·998	880·998	24·475	70·53	0·10	1·04
4	7839·973	8720·972	33·371	76·19	0·05	1·03
5	77607·687	86328·656	42·270	80·05	0·03	1·03
6	768236·376	854565·001	51·169	82·83	0·02	1·03
		$\mu = 9\cdot00000$				
1	9·000	10·000	8·000	45·00	0·87	3·56
2	99·000	109·000	17·636	61·93	0·21	1·17
3	1079·998	1188·998	27·525	70·96	0·09	1·03
4	11780·972	12969·970	37·432	76·57	0·05	1·02
5	128510·484	141480·437	47·340	80·37	0·03	1·03
6	1401834·252	1543314·503	57·248	83·12	0·02	1·03
		$\mu = 10\cdot00000$				
1	10·000	11·000	9·000	45·45	0·85	3·47
2	120·000	131·000	19·667	62·34	0·20	1·15
3	1429·997	1560·997	30·566	71·32	0·09	1·03
4	17039·957	18600·953 .	41·481	76·88	0·05	1·02
5	203049·375	221650·312	52·397	80·64	0·03	1·03
6	2419551·005	2641201·006	63·313	83·35	0·02	1·03

Table 3.9.0 *Stability functions (p positive)*
$$(p = P/P_E)$$

p	S	C	S^*	p	S	C	S^*
0·00	4·0000	0·5000	3·0000	0·50	3·2945	0·6659	1·8338
0·01	3·9870	0·5025	2·9803	0·51	3·2793	0·6703	1·8057
0·02	3·9737	0·5050	2·9603	0·52	3·2640	0·6749	1·7774
0·03	3·9604	0·5075	2·9403	0·53	3·2487	0·6795	1·7488
0·04	3·9471	0·5101	2·9202	0·54	3·2334	0·6841	1·7200
0·05	3·9338	0·5127	2·8999	0·55	3·2180	0·6889	1·6909
0·06	3·9204	0·5153	2·8795	0·56	3·2025	0·6937	1·6615
0·07	3·9071	0·5179	2·8590	0·57	3·1870	0·6985	1·6319
0·08	3·8936	0·5206	2·8384	0·58	3·1715	0·7035	1·6020
0·09	3·8802	0·5233	2·8177	0·59	3·1559	0·7085	1·5718
0·10	3·8667	0·5260	2·7968	0·60	3·1403	0·7136	1·5414
0·11	3·8531	0·5288	2·7758	0·61	3·1246	0·7187	1·5106
0·12	3·8396	0·5316	2·7547	0·62	3·1088	0·7239	1·4795
0·13	3·8260	0·5344	2·7334	0·63	3·0930	0·7292	1·4482
0·14	3·8123	0·5372	2·7120	0·64	3·0771	0·7346	1·4165
0·15	3·7987	0·5401	2·6905	0·65	3·0612	0·7401	1·3845
0·16	3·7850	0·5430	2·6688	0·66	3·0453	0·7456	1·3522
0·17	3·7712	0·5460	2·6470	0·67	3·0293	0·7513	1·3196
0·18	3·7574	0·5490	2·6251	0·68	3·0132	0·7570	1·2866
0·19	3·7436	0·5520	2·6030	0·69	2·9971	0·7628	1·2533
0·20	3·7297	0·5550	2·5808	0·70	2·9809	0·7687	1·2197
0·21	3·7158	0·5581	2·5584	0·71	2·9647	0·7746	1·1856
0·22	3·7019	0·5612	2·5359	0·72	2·9484	0·7807	1·1513
0·23	3·6879	0·5644	2·5132	0·73	2·9320	0·7869	1·1165
0·24	3·6739	0·5676	2·4904	0·74	2·9156	0·7932	1·0814
0·25	3·6598	0·5708	2·4674	0·75	2·8991	0·7995	1·0458
0·26	3·6457	0·5741	2·4443	0·76	2·8826	0·8060	1·0099
0·27	3·6315	0·5774	2·4210	0·77	2·8660	0·8126	0·9736
0·28	3·6174	0·5807	2·3975	0·78	2·8494	0·8193	0·9368
0·29	3·6031	0·5841	2·3738	0·79	2·8327	0·8261	0·8997
0·30	3·5889	0·5875	2·3500	0·80	2·8159	0·8330	0·8621
0·31	3·5746	0·5910	2·3261	0·81	2·7991	0·8400	0·8240
0·32	3·5602	0·5945	2·3019	0·82	2·7822	0·8472	0·7855
0·33	3·5458	0·5981	2·2776	0·83	2·7653	0·8544	0·7465
0·34	3·5314	0·6017	2·2531	0·84	2·7483	0·8618	0·7071
0·35	3·5169	0·6053	2·2284	0·85	2·7312	0·8693	0·6671
0·36	3·5024	0·6090	2·2035	0·86	2·7141	0·8770	0·6267
0·37	3·4878	0·6127	2·1784	0·87	2·6969	0·8848	0·5857
0·38	3·4732	0·6165	2·1532	0·88	2·6797	0·8927	0·5442
0·39	3·4586	0·6203	2·1277	0·89	2·6623	0·9008	0·5022
0·40	3·4439	0·6242	2·1021	0·90	2·6450	0·9090	0·4596
0·41	3·4292	0·6281	2·0762	0·91	2·6275	0·9173	0·4165
0·42	3·4144	0·6321	2·0502	0·92	2·6100	0·9258	0·3727
0·43	3·3996	0·6361	2·0239	0·93	2·5924	0·9345	0·3284
0·44	3·3847	0·6402	1·9974	0·94	2·5748	0·9433	0·2835
0·45	3·3698	0·6443	1·9707	0·95	2·5570	0·9523	0·2379
0·46	3·3548	0·6485	1·9438	0·96	2·5393	0·9615	0·1917
0·47	3·3398	0·6528	1·9166	0·97	2·5214	0·9709	0·1448
0·48	3·3247	0·6571	1·8893	0·98	2·5035	0·9804	0·0972
0·49	3·3096	0·6614	1·8617	0·99	2·4855	0·9901	0·0490

Table 3.9.0 (Cont.)

p	S	C	S^*	p	S	C	S^*
1·00	2·4674	1·0000	0·0000	1·50	1·4570	1·9731	−4·2148
1·01	2·4493	1·0101	−0·0496	1·51	1·4342	2·0114	−4·3681
1·02	2·4311	1·0204	−0·1001	1·52	1·4114	2·0512	−4·5268
1·03	2·4128	1·0309	−0·1514	1·53	1·3884	2·0926	−4·6911
1·04	2·3944	1·0416	−0·2034	1·54	1·3653	2·1356	−4·8614
1·05	2·3760	1·0526	−0·2563	1·55	1·3420	2·1804	−5·0381
1·06	2·3575	1·0638	−0·3101	1·56	1·3187	2·2270	−5·2215
1·07	2·3389	1·0752	−0·3647	1·57	1·2952	2·2757	−5·4121
1·08	2·3202	1·0868	−0·4203	1·58	1·2716	2·3264	−5·6104
1·09	2·3015	1·0987	−0·4767	1·59	1·2479	2·3794	−5·8168
1·10	2·2827	1·1109	−0·5342	1·60	1·2240	2·4348	−6·0318
1·11	2·2638	1·1233	−0·5926	1·61	1·2000	2·4927	−6·2562
1·12	2·2448	1·1360	−0·6521	1·62	1·1759	2·5534	−6·4904
1·13	2·2258	1·1490	−0·7126	1·63	1·1516	2·6170	−6·7353
1·14	2·2066	1·1623	−0·7742	1·64	1·1272	2·6838	−6·9917
1·15	2·1874	1·1759	−0·8369	1·65	1·1027	2·7540	−7·2603
1·16	2·1681	1·1898	−0·9008	1·66	1·0780	2·8278	−7·5422
1·17	2·1488	1·2040	−0·9659	1·67	1·0532	2·9056	−7·8384
1·18	2·1293	1·2185	−1·0323	1·68	1·0282	2·9877	−8·1500
1·19	2·1097	1·2335	−1·0999	1·69	1·0031	3·0744	−8·4785
1·20	2·0901	1·2487	−1·1689	1·70	0·9779	3·1662	−8·8251
1·21	2·0704	1·2644	−1·2393	1·71	0·9525	3·2634	−9·1916
1·22	2·0506	1·2804	−1·3110	1·72	0·9270	3·3667	−9·5798
1·23	2·0307	1·2968	−1·3843	1·73	0·9013	3·4765	−9·9917
1·24	2·0107	1·3137	−1·4591	1·74	0·8754	3·5936	−10·4297
1·25	1·9906	1·3309	−1·5355	1·75	0·8494	3·7186	−10·8963
1·26	1·9705	1·3487	−1·6135	1·76	0·8233	3·8524	−11·3947
1·27	1·9502	1·3669	−1·6933	1·77	0·7969	3·9959	−11·9282
1·28	1·9299	1·3855	−1·7748	1·78	0·7705	4·1503	−12·5008
1·29	1·9095	1·4047	−1·8582	1·79	0·7438	4·3168	−13·1171
1·30	1·8889	1·4244	−1·9435	1·80	0·7170	4·4969	−13·7825
1·31	1·8683	1·4447	−2·0309	1·81	0·6901	4·6923	−14·5031
1·32	1·8476	1·4655	−2·1203	1·82	0·6629	4·9051	−15·2864
1·33	1·8268	1·4869	−2·2118	1·83	0·6356	5·1376	−16·1409
1·34	1·8058	1·5089	−2·3057	1·84	0·6081	5·3928	−17·0772
1·35	1·7848	1·5316	−2·4019	1·85	0·5805	5·6741	−18·1075
1·36	1·7637	1·5549	−2·5005	1·86	0·5526	5·9858	−19·2474
1·37	1·7425	1·5790	−2·6017	1·87	0·5246	6·3330	−20·5154
1·38	1·7212	1·6038	−2·7056	1·88	0·4964	6·7222	−21·9346
1·39	1·6997	1·6293	−2·8123	1·89	0·4680	7·1614	−23·5341
1·40	1·6782	1·6557	−2·9220	1·90	0·4394	7·6610	−25·3514
1·41	1·6566	1·6828	−3·0347	1·91	0·4107	8·2342	−27·4339
1·42	1·6348	1·7109	−3·1506	1·92	0·3817	8·8988	−29·8457
1·43	1·6130	1·7399	−3·2699	1·93	0·3526	9·6782	−32·6713
1·44	1·5910	1·7699	−3·3928	1·94	0·3232	10·6052	−36·0288
1·45	1·5690	1·8009	−3·5193	1·95	0·2937	11·7259	−40·0843
1·46	1·5468	1·8329	−3·6497	1·96	0·2639	13·1082	−45·0823
1·47	1·5245	1·8661	−3·7843	1·97	0·2340	14·8555	−51·3959
1·48	1·5021	1·9005	−3·9232	1·98	0·2038	17·1347	−59·6261
1·49	1·4796	1·9361	−4·0666	1·99	0·1734	20·2317	−70·8029

Table 3.9.0 (Cont.)

p	S	C	S^*	p	S	C	S^*
2·00	0·1428	24·6820	−86·8567	2·50	−1·7497	−2·6732	10·7546
2·01	0·1120	31·6235	−111·8868	2·51	−1·7972	−2·6202	10·5427
2·02	0·0810	43·9555	−156·3408	2·52	−1·8452	−2·5695	10·3389
2·03	0·0497	71·9438	−257·2098	2·53	−1·8937	−2·5210	10·1428
2·04	0·0182	197·2413	−708·7170	2·54	−1·9428	−2·4744	9·9537
2·05	−0·0134	−267·5336	966·0068	2·55	−1·9924	−2·4297	9·7715
2·06	−0·0453	−79·8414	289·6704	2·56	−2·0425	−2·3869	9·5955
2·07	−0·0775	−46·9723	171·2140	2·57	−2·0932	−2·3457	9·4254
2·08	−0·1100	−33·2971	121·9198	2·58	−2·1445	−2·3060	9·2610
2·09	−0·1426	−25·8047	94·9040	2·59	−2·1963	−2·2679	9·1019
2·10	−0·1756	−21·0744	77·8408	2·60	−2·2488	−2·2312	8·9478
2·11	−0·2087	−17·8171	66·0855	2·61	−2·3018	−2·1958	8·7984
2·12	−0·2422	−15·4374	57·4925	2·62	−2·3555	−2·1618	8·6536
2·13	−0·2759	−13·6227	50·9352	2·63	−2·4098	−2·1289	8·5129
2·14	−0·3098	−12·1932	45·7659	2·64	−2·4648	−2·0971	8·3763
2·15	−0·3440	−11·0383	41·5859	2·65	−2·5204	−2·0665	8·2436
2·16	−0·3785	−10·0856	38·1344	2·66	−2·5767	−2·0368	8·1145
2·17	−0·4133	−9·2863	35·2359	2·67	−2·6336	−2·0082	7·9888
2·18	−0·4483	−8·6063	32·7669	2·68	−2·6913	−1·9805	7·8665
2·19	−0·4837	−8·0206	30·6380	2·69	−2·7497	−1·9537	7·7472
2·20	−0·5193	−7·5110	28·7827	2·70	−2·8088	−1·9278	7·6310
2·21	−0·5552	−7·0635	27·1515	2·71	−2·8687	−1·9026	7·5176
2·22	−0·5914	−6·6675	25·7057	2·72	−2·9294	−1·8783	7·4069
2·23	−0·6279	−6·3146	24·4149	2·73	−2·9908	−1·8547	7·2989
2·24	−0·6648	−5·9980	23·2554	2·74	−3·0531	−1·8318	7·1933
2·25	−0·7019	−5·7126	22·2076	2·75	−3·1162	−1·8096	7·0901
2·26	−0·7394	−5·4538	21·2562	2·76	−3·1801	−1·7881	6·9891
2·27	−0·7771	−5·2182	20·3880	2·77	−3·2449	−1·7672	6·8903
2·28	−0·8153	−5·0028	19·5925	2·78	−3·3105	−1·7469	6·7936
2·29	−0·8537	−4·8051	18·8607	2·79	−3·3771	−1·7272	6·6989
2·30	−0·8925	−4·6230	18·1852	2·80	−3·4446	−1·7080	6·6061
2·31	−0·9316	−4·4548	17·5595	2·81	−3·5131	−1·6894	6·5151
2·32	−0·9711	−4·2989	16·9782	2·82	−3·5825	−1·6713	6·4259
2·33	−1·0110	−4·1540	16·4365	2·83	−3·6529	−1·6537	6·3383
2·34	−1·0512	−4·0191	15·9304	2·84	−3·7244	−1·6366	6·2524
2·35	−1·0917	−3·8930	15·4564	2·85	−3·7969	−1·6199	6·1680
2·36	−1·1327	−3·7750	15·0114	2·86	−3·8705	−1·6037	6·0851
2·37	−1·1740	−3·6644	14·5928	2·87	−3·9452	−1·5879	6·0037
2·38	−1·2158	−3·5605	14·1982	2·88	−4·0210	−1·5725	5·9236
2·39	−1·2579	−3·4626	13·8255	2·89	−4·0980	−1·5575	5·8448
2·40	−1·3004	−3·3703	13·4727	2·90	−4·1762	−1·5429	5·7673
2·41	−1·3434	−3·2831	13·1384	2·91	−4·2556	−1·5287	5·6911
2·42	−1·3867	−3·2007	12·8209	2·92	−4·3363	−1·5149	5·6160
2·43	−1·4305	−3·1225	12·5190	2·93	−4·4182	−1·5013	5·5420
2·44	−1·4747	−3·0484	12·2314	2·94	−4·5015	−1·4882	5·4692
2·45	−1·5194	−2·9780	11·9572	2·95	−4·5862	−1·4753	5·3974
2·46	−1·5645	−2·9111	11·6953	2·96	−4·6723	−1·4628	5·3266
2·47	−1·6101	−2·8473	11·4448	2·97	−4·7599	−1·4506	5·2568
2·48	−1·6561	−2·7865	11·2050	2·98	−4·8489	−1·4386	5·1880
2·49	−1·7027	−2·7285	10·9752	2·99	−4·9395	−1·4270	5·1200

Table 3.9.0 (*Cont.*)

p	S	C	S^*	p	S	C	S^*
3·00	−5·0316	−1·4156	5·0530	3·50	−13·7175	−1·0823	2·3534
3·01	−5·1254	−1·4045	4·9868	3·51	−14·0586	−1·0788	2·3063
3·02	−5·2208	−1·3937	4·9214	3·52	−14·4133	−1·0754	2·2593
3·03	−5·3180	−1·3831	4·8567	3·53	−14·7825	−1·0721	2·2123
3·04	−5·4170	−1·3728	4·7929	3·54	−15·1672	−1·0689	2·1655
3·05	−5·5178	−1·3627	4·7298	3·55	−15·5683	−1·0658	2·1187
3·06	−5·6205	−1·3528	4·6674	3·56	−15·9870	−1·0627	2·0721
3·07	−5·7252	−1·3432	4·6056	3·57	−16·4246	−1·0598	2·0255
3·08	−5·8318	−1·3338	4·5446	3·58	−16·8824	−1·0569	1·9789
3·09	−5·9406	−1·3246	4·4841	3·59	−17·3617	−1·0541	1·9324
3·10	−6·0515	−1·3156	4·4243	3·60	−17·8645	−1·0514	1·8859
3·11	−6·1646	−1·3068	4·3651	3·61	−18·3920	−1·0487	1·8395
3·12	−6·2801	−1·2983	4·3065	3·62	−18·9468	−1·0462	1·7932
3·13	−6·3979	−1·2899	4·2484	3·63	−19·5307	−1·0437	1·7468
3·14	−6·5182	−1·2817	4·1908	3·64	−20·1463	−1·0412	1·7005
3·15	−6·6410	−1·2737	4·1338	3·65	−20·7961	−1·0389	1·6542
3·16	−6·7664	−1·2658	4·0773	3·66	−21·4834	−1·0366	1·6079
3·17	−6·8946	−1·2582	4·0212	3·67	−22·2115	−1·0345	1·5616
3·18	−7·0256	−1·2507	3·9656	3·68	−22·9840	−1·0323	1·5153
3·19	−7·1596	−1·2434	3·9105	3·69	−23·8056	−1·0303	1·4690
3·20	−7·2966	−1·2362	3·8557	3·70	−24·6809	−1·0283	1·4227
3·21	−7·4367	−1·2292	3·8014	3·71	−25·6155	−1·0264	1·3764
3·22	−7·5801	−1·2224	3·7476	3·72	−26·6158	−1·0246	1·3301
3·23	−7·7270	−1·2157	3·6941	3·73	−27·6891	−1·0228	1·2837
3·24	−7·8773	−1·2091	3·6409	3·74	−28·8439	−1·0211	1·2373
3·25	−8·0314	−1·2027	3·5882	3·75	−30·0896	−1·0195	1·1908
3·26	−8·1892	−1·1965	3·5357	3·76	−31·4380	−1·0179	1·1443
3·27	−8·3511	−1·1903	3·4836	3·77	−32·9022	−1·0164	1·0978
3·28	−8·5171	−1·1844	3·4319	3·78	−34·4983	−1·0150	1·0511
3·29	−8·6874	−1·1785	3·3804	3·79	−36·2449	−1·0137	1·0045
3·30	−8·8622	−1·1728	3·3293	3·80	−38·1643	−1·0124	0·9578
3·31	−9·0417	−1·1672	3·2784	3·81	−40·2845	−1·0111	0·9109
3·32	−9·2261	−1·1617	3·2278	3·82	−42·6381	−1·0100	0·8640
3·33	−9·4156	−1·1564	3·1775	3·83	−45·2670	−1·0089	0·8171
3·34	−9·6105	−1·1512	3·1274	3·84	−48·2220	−1·0079	0·7700
3·35	−9·8110	−1·1461	3·0776	3·85	−51·5692	−1·0069	0·7229
3·36	−10·0174	−1·1411	3·0280	3·86	−55·3916	−1·0060	0·6756
3·37	−10·2299	−1·1362	2·9787	3·87	−59·8005	−1·0051	0·6282
3·38	−10·4488	−1·1314	2·9295	3·88	−64·9404	−1·0044	0·5807
3·39	−10·6745	−1·1268	2·8806	3·89	−71·0114	−1·0036	0·5331
3·40	−10·9072	−1·1222	2·8318	3·90	−78·2933	−1·0030	0·4854
3·41	−11·1474	−1·1178	2·7833	3·91	−87·1887	−1·0024	0·4376
3·42	−11·3954	−1·1135	2·7349	3·92	−98·3003	−1·0019	0·3896
3·43	−11·6517	−1·1092	2·6867	3·93	−112·5861	−1·0014	0·3415
3·44	−11·9166	−1·1051	2·6387	3·94	−131·6166	−1·0010	0·2932
3·45	−12·1907	−1·1010	2·5908	3·95	−158·2493	−1·0007	0·2447
3·46	−12·4744	−1·0971	2·5430	3·96	−198·1642	−1·0004	0·1962
3·47	−12·7684	−1·0933	2·4954	3·97	−264·6404	−1·0002	0·1475
3·48	−13·0731	−1·0895	2·4480	3·98	−397·3864	−1·0000	0·0985
3·49	−13·3892	−1·0859	2·4006	3·99	−794·1709	−1·0000	0·0494
				4·00	− ∞	−1·0000	0·0000

Table 3.9.1 *Stability functions (p negative)*
$$(p = P/P_E).$$

p	S	C	S^*	p	S	C	S^*
0·00	4·0000	0·5000	3·0000	−2·00	6·1468	0·2600	5·7313
−0·05	4·0654	0·4880	3·0973	−2·05	6·1916	0·2572	5·7821
−0·10	4·1299	0·4765	3·1920	−2·10	6·2360	0·2544	5·8324
−0·15	4·1937	0·4657	3·2843	−2·15	6·2801	0·2517	5·8823
−0·20	4·2566	0·4553	3·3743	−2·20	6·3239	0·2491	5·9316
−0·25	4·3189	0·4454	3·4621	−2·25	6·3675	0·2465	5·9805
−0·30	4·3804	0·4360	3·5478	−2·30	6·4107	0·2440	6·0290
−0·35	4·4412	0·4269	3·6317	−2·35	6·4536	0·2416	6·0770
−0·40	4·5012	0·4183	3·7136	−2·40	6·4963	0·2392	6·1246
−0·45	4·5607	0·4100	3·7939	−2·45	6·5387	0·2369	6·1718
−0·50	4·6194	0·4021	3·8725	−2·50	6·5808	0·2346	6·2186
−0·55	4·6776	0·3945	3·9495	−2·55	6·6226	0·2324	6·2650
−0·60	4·7351	0·3872	4·0251	−2·60	6·6642	0·2302	6·3110
−0·65	4·7920	0·3802	4·0992	−2·65	6·7055	0·2281	6·3566
−0·70	4·8483	0·3735	4·1720	−2·70	6·7466	0·2260	6·4019
−0·75	4·9041	0·3670	4·2435	−2·75	6·7874	0·2240	6·4468
−0·80	4·9593	0·3608	4·3137	−2·80	6·8279	0·2220	6·4914
−0·85	5·0139	0·3548	4·3828	−2·85	6·8683	0·2201	6·5356
−0·90	5·0681	0·3490	4·4507	−2·90	6·9084	0·2182	6·5795
−0·95	5·1217	0·3434	4·5176	−2·95	6·9482	0·2163	6·6230
−1·00	5·1748	0·3381	4·5834	−3·00	6·9878	0·2145	6·6662
−1·05	5·2274	0·3329	4·6482	−3·05	7·0272	0·2127	6·7092
−1·10	5·2795	0·3279	4·7120	−3·10	7·0664	0·2110	6·7518
−1·15	5·3312	0·3230	4·7750	−3·15	7·1053	0·2093	6·7941
−1·20	5·3824	0·3183	4·8370	−3·20	7·1441	0·2076	6·8361
−1·25	5·4332	0·3138	4·8982	−3·25	7·1826	0·2060	6·8778
−1·30	5·4835	0·3094	4·9586	−3·30	7·2209	0·2044	6·9192
−1·35	5·5334	0·3051	5·0182	−3·35	7·2590	0·2028	6·9604
−1·40	5·5828	0·3010	5·0770	−3·40	7·2969	0·2013	7·0013
−1·45	5·6319	0·2970	5·1351	−3·45	7·3346	0·1998	7·0419
−1·50	5·6806	0·2931	5·1924	−3·50	7·3721	0·1983	7·0823
−1·55	5·7288	0·2894	5·2491	−3·55	7·4094	0·1968	7·1224
−1·60	5·7767	0·2857	5·3051	−3·60	7·4465	0·1954	7·1622
−1·65	5·8242	0·2822	5·3604	−3·65	7·4834	0·1940	7·2018
−1·70	5·8713	0·2787	5·4151	−3·70	7·5201	0·1926	7·2412
−1·75	5·9181	0·2754	5·4693	−3·75	7·5567	0·1912	7·2803
−1·80	5·9645	0·2721	5·5228	−3·80	7·5930	0·1899	7·3192
−1·85	6·0106	0·2690	5·5757	−3·85	7·6292	0·1886	7·3578
−1·90	6·0563	0·2659	5·6281	−3·90	7·6652	0·1873	7·3962
−1·95	6·1017	0·2629	5·6800	−3·95	7·7011	0·1861	7·4344
				−4·00	7·7367	0·1848	7·4724

INDEX

A section area, 109
A   coefficient matrix, 93
algebraic equations, 180; over-deter-mined, 181
analysis, 90, 184, 196
approximation, 169, 181; moment definit-ion, 171
arbitrary constants, interpretation, 190; beam, 15, 23; column, 114

beam, 11; composite, 160; deep, 176; section stiffness ($ß = EI$), 7, 23
bending moment (B.M.) $\longrightarrow$ notation, 1
boundary conditions, 2; beam, 23; column, 115; composite, 162
boundary value type problems (B.V.T.), 6, 7
bracing, cross, 157
bridge, decks, 142; structure, 37

C   coefficient matrix, 95
cable, 153, 163
carryover factor, 34, 130
checks on working, 42
choice of origin, 32, 68
classical theory, xiii
collapse load, 17, 21, 31, 43, 54, 71, 79, 83
column, 112; boundary condition, 115; continuous, 127, 132; discontinuous particular integrals, 121, 124; Euler, 116; interpretation of constants, 115
complementary function (C.F.), 2
composite beam, 175
compressible shear wall, 108
computation, machine, xi, 55, 167; manual, xi, 167
concordant profile, 156
conditions for zero constants, 118
constants, arbitrary, beam, 15; column, 115
continuity, 1
continuous, xii
contragredience, 201
contravariant vector, 201
couple stress, 202
covariant vector, 201
cross bracing, 156
curvature, 12, 22

$\mathcal{D}$ an elastic constant, 175
deficient, 49

design, 184, 196
determinate, 24, 49, 113
diagonal matrix, 188, 200
differential equation, xii, 1, 7; order, 2, 7, 23, 177; ordinary (O.D.E.), 1
dimensional analysis, 74
dimensionless numbers, 74
Dirac delta function, 9, 10
discontinuity, 1; beam, 15, 25; displace-ment, 28; end point, 31; rotation, 144; simultaneous, 32; symbol for [ ], 4
discontinuous, xii; particular integral (D.P.I.), 4; for beam, 15, 25; for column, 125
displaced form, xii
displacement, xi; axial, 152; disconti-nuity, 28
dissipative material, 199
distribution factors, 79
duality, 14, 18
ductility, 13

E Young's modulus, 23
eigenvalue, 200; problem, 116; vector, 116
elastic, 11; centre, 193; foundation, 6, 126; plastic, 11, 21; strain-hardening, 11
energy, work, xiii, 19, 201
Euler column, 116
evenness, 7, 26, 42

F shear force, 13
fixed end moments, 79
force, notation $\rightarrow$, 1; at joints, 70; dis-continuity, 16; force/displacement rela-tion, 22
formulation of problem, 31
framework, 47; mill-type, 58; multi-storey, 80, 90; two-storey, 72, 75
foundations of beam theory, 169

G shear modulus, 176
$\mathcal{G}$ torsional stiffness, 145
global property, 21, 48, 50, 181
governing differential equation (G.D.E.), 7
grillage, 143

Heaviside step function, 9, 10
hinge, plastic, xii, 14, 28
homogeneous problem, 115

I statical indeterminacy index, 48
indeterminate, 24, 48
inertia, second moment of area (I), 23
inextensibility, 63; beam, 108; cable, 163
initial stress, 152
initial value type problem (I.V.T.), 6
instability, 112, 132
interpretation of integration constants, beam, 24; column, 115

joint, 56

kinematic freedoms, 49, 50, 63

L typical span, 16
$L(..)$ typical differential operator, 2, 19, 179
Laplace Transform (L.T.), 6
least squares, 181
limit theorems, 19
linear profile, 156
load factor, 21, 79
local property, 21, 48, 50
lower bound, 20

M internal bending moment, 11
Macaulay bracket, xii
machine computation, 167
manual computation, 167
material properties, 11, 14, 22
matrix, diagonal, 188, 200; orthogonal, 200; positive definite, 200; rank, 18; similar, 201; symmetric, 187
maximum–minimum stiffness theorem, 185
mechanism, 11, 14, 19
models, 74
Mohr circle, 188, 192, 195, 201
moment, bending, xi, 13; definition, 170; physical assumption, 171

necessary and sufficient conditions, 183
nonconservative, 199
nonhomogeneous problem, 119, 142
nonorthogonal axes, 186
notation, see symbols (principal)
N.T.S. not to scale, 166

operational methods (O.M.), 6
order of differential equation, 2
ordinary differential equation (O.D.E.), 1
origin, choice of, 32, 68
orthogonal matrix, 200
over-determined systems, 181

P axial force, 114

p distributed loading, 13
P_n shear wall function, 97
particular integral (P.I.), 3; continuous (C.P.I.), 3; discontinuous (D.P.I.), 3
physical interpretation of integration constants, beam, 24; column, 115
plastic hinge, xii, 14, 28
prestress, 153
principal, 185; axes, 191, 196; property, 184
proportional response, 11

Q_n shear wall function, 97

rank of matrix, 18
rational functions, 74
redundant, *see* indeterminate
rigid–plastic, 11
rings, topological, 48

S shear wall force reduction factor, 104
safety factor, 21
secant strut formula, 121
second moment of area (I), 23
second order beam theory, 176
self-adjoint, 19, 199
sequence, 1
shape factor, 13
shear centre, 196; connexion, 160; deflexion, 176; force (F), 8, 24; wall, 91
sign conventions, 16
similarity transformation, 201
simultaneous discontinuity, 32
single bay multi-storey frame, 91
slenderness ratio, 118
slip, 160
Southwell plot, 139
space frame, 148
span (L), 16; ß beam section stiffness (EI), 7, 23
stability functions, 128, 130
static check, 88
statical indeterminacy, 24, 48
stiffness (ß), 7, 11, 21, 23; factor, 34, 130
strain hardening, 12
strength, 11, 21
sway, 48, 142, 157
symbols (principal), ß section stiffness (EI), 7, 23; E Young's modulus, 23; I second moment of area, 23; L typical span, 16; M internal bending moment, 11; w transverse displacement, 14; x independent length variable, 2
symmetry, 26, 42

Index

temperature stresses, 152
tensor, cartesian, 172
time-like quality, 42
topology of structure, 48
transverse displacement (w), 14
twisting, 143

upper bound, 20

$\mathscr{V}$ wall vertical force, 100
variation of parameters, 3

w displacement vector, 93, 108, 109
w transverse displacement, 14
$\mathscr{W}$ shear wall stiffness, 92
work, energy, xiii, 19, 201
Wronskian, 4

x independent length variable, 2

yield, 19; moment, 13
Young's modulus (E), 23